A Postphenomenological Inquiry of Cell Phones

Postphenomenology and the Philosophy of Technology

Technological advances affect everything from our understandings of ethics, politics, and communication to gender, science, and selfhood. Philosophical reflection on technology helps draw out and analyze the nature of these changes, and helps us understand both the broad patterns and the concrete details of technological effects. This book series provides a publication outlet for the field of the philosophy of technology in general, and the school of thought called "postphenomenology" in particular. Philosophy of technology applies insights from the history of philosophy to current issues in technology, and reflects on how technological developments change our understanding of philosophical issues. In response, postphenomenology analyzes human relationships with technologies, while integrating philosophical commitments of the American pragmatist tradition of thought.

A Postphenomenological Inquiry of Cell Phones: Genealogies, Meanings, and Becoming by Galit P. Wellner
Technoscience and Postphenomenology: The Manhattan Papers, edited by Jan Kyrre Berg O. Friis and Robert P. Crease
Postphenomenological Investigations: Essays on Human–Technology Relations, edited by Peter-Paul Verbeek and Robert Rosenberger
Design, Meditation, and the Posthuman, edited by Dennis M. Weiss, Amy D. Propen, and Colbey Emmerson Reid

A Postphenomenological Inquiry of Cell Phones

Genealogies, Meanings, and Becoming

Galit P. Wellner

LEXINGTON BOOKS
Lanham • Boulder • New York • London

Published by Lexington Books
An imprint of The Rowman & Littlefield Publishing Group, Inc.
4501 Forbes Boulevard, Suite 200, Lanham, Maryland 20706
www.rowman.com

Unit A, Whitacre Mews, 26-34 Stannary Street, London SE11 4AB

British Library Cataloguing in Publication Information Available

Library of Congress Cataloging-in-Publication Data

Names: Wellner, Galit P., author.
Title: A postphenomenological inquiry of cell phones : genealogies, meanings, and becoming/by Galit P. Wellner.
Description: Lanham : Lexington Books, [2016] | Series: Postphenomenology and the philosophy of technology | Includes bibliographical references and index.
Identifiers: LCCN 2015033448| ISBN 9780739198483 (cloth : alk. paper) | ISBN 9780739198490 (electronic)
Subjects: LCSH: Mobile computing—Philosophy. | Cell phones—Philosophy. | Phenomenological sociology.
Classification: LCC QA76.59 .W46 2016 | DDC 004—dc23 LC record available at http://lccn.loc.gov/2015033448

∞™ The paper used in this publication meets the minimum requirements of American National Standard for Information Sciences—Permanence of Paper for Printed Library Materials, ANSI/NISO Z39.48-1992.

Printed in the United States of America

Contents

List of Illustrations

FIGURES

TABLES

Acknowledgments

This book is based on research conducted as part of my PhD degree. I am deeply grateful to my dissertation advisors Don Ihde and Lyat Friedman for their support and thoughtful guidance that they have continued to provide me until today. It was a great honor to be their "advisee," as Don terms this status. They taught me how to think and then how to write in my own voice. These are important assets I acquired for my academic pursuits. Their faith in my ideas gave me the confidence to publish at relatively early stages in my studies, and this book could not have been written without them.

I also owe a big intellectual debt to the Postphenomenology Group, who received me with open arms. I quickly felt part of this warm and intelligent academic family whose "father" is with no doubt Don. Postphenomenologists have played an important role in the development of my ideas and later in refining them. There are a few people that I met thanks to this group and who deserve special mention. I want to thank Peter-Paul Verbeek, Ciano Aydin, Yoni Van den Eede and Pieter Lemmens for their careful reading, invaluably constructive comments and warm encouragement. I also want to thank Robert Rosenberger, Stacey Irwin, Lars Botin, Bob Scharff, Tamar Sharon and Marie-Christine Nizzi for the illuminating and insightful discussions. Thank you all for becoming my friends and for accompanying me during the long writing process.

I would like to thank the STS program at Bar-Ilan University and the Department of Education at Ben-Gurion University of the Negev for their financial support throughout the writing process. Specifically, I am grateful to Oren Harman and Roni Aviram whose aid made things easier.

Some of the chapters appearing in this book are reworked versions of previously published articles. Thankful acknowledgment is made to the following publishing houses and editors for their kind permission to republish the following articles: "Wall-Window-Screen: How the Cell Phone Mediates a Worldview for Us." *Humanities and Technology Review* 30 (2011): 77–104; "No Longer a Phone: The Cellphone as an Enabler of Augmented Reality." *Transfers* 3.2 (2013): 70–88; "The Quasi-Face of the Cell Phone: Rethinking Alterity and Screens." *Human Studies* (2014): 1–18.

A very special thank you goes to my dearest partner Jacob, my beloved children Jonathan, Naama and Oded, and to my whole family, who have been supportive throughout the years of this writing project. It was important to know that I have a solid support system. Specifically, I am in debt to my English-speaking family members, Rachia Wellner, Beatrice and Micky Gilat, Vivian Mishan, David and Stan Robins, for editing many of the texts that have become part of this book, and for always helping me improve my second-language capabilities. I would like to express my gratitude not only to my family members who assisted me in the last years, but also to my friends Odelia Rosin, Yael Rozin, Rachel Even, Ady Wiernik, Talia Fried and Amira Cohen who gave me constructive comments and warm encouragement. I appreciate their feedbacks, insights and friendship. Last but not least, I wish to thank my copy editor Michelle Spektor for her good comments and patience.

Galit P. Wellner
Tel Aviv, May 2015

Introduction

WHY STUDY THE CELL PHONE?

Cell phones are now everywhere; in the streets of New York and Nairobi, in meeting rooms and classrooms, in theater halls and at funerals, in the bedroom, in the kitchen, in the washroom—everywhere. This technological artifact spans across continents, usages, ages and has now become as common as clothing, dwellings and cooking, to name a few of the elementary technologies that accompany human beings no matter where they are. After a few decades of constant penetration to each and every aspect of our lives, it is now rare to meet a person who does not use a cell phone.

The cell phone becomes an important participant in our everydayness, assisting us not only to communicate with others as its name ("phone") suggests, but also to record where we go, when we go and how; to entertain us with music, books and videos; to help us with a flashlight, a compass and a scratchpad, and more. That is why people find it difficult to function without their cell phones. Many would return home if they forget it, but only few would return for their wallet. We have become addicted to our cell phones. A historical account can describe how we reached this point, but can hardly explain why this technology is so addictive across cultures, ages and genders. Fetish and magic can provide some justifications but most people are reluctant to identify themselves in these terms. This is where philosophy of technology, and especially a branch of it known as postphenomenology, can provide reasoning.

This book consists of a postphenomenological account of the cell phone. It describes the major stages of its evolution and the characteristics that remain active and relevant throughout these stages. But before approaching this analysis, I would like to ponder the history of technology and place the cell phone in a historical context of technological evolution. My underlying assumption is that cell phones are emblematic for new types of technologies that I call "digital technologies" (DTs). These technologies are operated on the digital sphere rather than the analog, and usually require digital production processes. They are dramatically different from tools and machines and maintain new relations with their users.

CONTEMPORARY TOOL ANALYSIS

In *Being Digital,* Nicholas Negroponte, the founding director of MIT's Media Lab, presents his vision for twenty-first-century technologies.[1] He maintains that technologies shift from moving atoms to moving bits. Moving atoms is characteristic of traditional technologies like printing, in which a physical book is transported from a publisher to a reader. In the twenty-first century, he claims, when we read a book electronically, bits are transferred (directly or indirectly) from the author to the reader. This digital delivery method transforms many of our everyday experiences into what Negroponte calls "digital life."

The insight that the twenty-first-century technologies are distinct is not limited to engineers. Sociologist Manuel Castells arrives at a similar conclusion. In *The Rise of the Network Society*,[2] the first volume of the trilogy *The Information Age: Economy, Society and Culture*, he asserts that "a technological revolution, centered around information technologies" took place toward the end of the second millennia.[3] According to Castells, the emergence of the Internet in the early 1990s, combined with political events such as the collapse of the communist states in Eastern Europe, led to a sociological change where "our societies are increasingly structured around a bipolar opposition between the Net and the self."[4] For Castells, the Internet is the "game changer" and a central component of the Information Age.[5]

Don Ihde, philosopher of technology, affixes the date of this shift in technology as early as the mid-twentieth century. The new technologies include not only information and communication technologies, as suggested by Negroponte and Castells, but also bio-, nano- and imaging technologies. He names the new breed of technologies "technoscience technologies" because they require scientific instrumentation in order to be produced and, in some cases, operated.[6] Compared to the early to mid-twentieth century, the period in which contemporary philosophy of technology emerged with philosophers like Ortega y Gasset and Martin Heidegger, "today's technologies evidence a quite different flavor from what was prominent during Heidegger's lifetime," Ihde argues.[7] The difference in flavor is exemplified in the shift from the dominancy of a hammer and a pen in our everydayness to the prevalence of computers and cell phones.[8] Let us have a "naïvely-realistic" look at technologies like the hammer. Ihde coins the term "naïve realism" to denote the "ordinary experience . . . where things are taken to be what they are seen to be."[9] A naïvely realistic look at the hammer reveals that holding it implies a functionality that is related to the here-and-now. It hints at physical work and is linked to the immediate environment of the workshop. Put differently, we feel "in place" when hammering nails with a hammer in the workshop, compared to visitors who enter the workshop with bare hands. Like the visitors of the workshop, we feel out-of-context if we walk down the street holding a hammer. At best, it may suggest we are on our way to nail something. The hammer (as a tool for nailing) has strong ties to the space of the workshop, and when placed in contexts outside of that space, can hardly carry such ties to it. Unlike the hammer that connects to a certain space and context, the cell phone disconnects us from the here-and-now, no matter where we are. At the same time, it partially transports us to a space that is distant in time and/or place, or a space that is virtual altogether. This is where we feel at home. Because it is not a physical space, we can feel at home everywhere. The naïvely realistic view exposes how cell phones produce a different "flavor" compared to hammers.

Negroponte, Castells and Ihde each conceptualize the interrelations between people and contemporary technologies in different ways. Though they each speak of digital, informational and technoscientific technologies, respectively, they all agree that these technologies have a different effect on us as individuals, as a society and as a culture. These technologies enframe our everyday experiences in a distinctive way. How did these technologies emerge? I have identified three phases in the evolution of technology: tools, machines, and DTs. This chapter provides a genealogical examination of each of these phases through the study of the interrelations between humans, technologies and environments. This interrelational analysis reveals that each generation of technology interacts mostly with a certain dominant human organ. The tool is typically operated by the hand, the machine by the body and the DT by the face. Consequently, each genealogical phase is characterized by certain technologies, a particular set of human practices, and a specific environment that serves as a background.

The Tool

The tool is a special object. Karl Marx in *Capital* defines a tool as "a thing, or a complex of thing[s], which the labourer interposes between himself and the subject of his labour, and which serves as the conductor of his activity."[10] In short, it is "an instrument of labour."[11] What turns an object into a tool is its participation in the labor process. Likewise, Heidegger conceives the tool as "something [that is] in order to . . . ,"[12] whose goal is labor, as in the case of the hammer which is for hammering. Tools, according to Heidegger, have a concrete function or utility. A hammer as well as a shoe and a clock are therefore defined by the "intention" associated with them. They are intended to be used for specific purposes such as labor.[13]

Marx further asserts that tools become what they are even before their participation in the labor process; they are made and produced with the intention of participating in this process. Tools are "specially prepared instruments"[14] in that an object becomes a tool if it is manufactured with the objective of creating a tool. The ability to plan in advance, Marx argues, enables humans to produce tools that are intended to be used for labor. This intention epitomizes the human ability to refer to the future. Marx demonstrates, "What distinguishes the worst architect from the best of bees is this, that the architect raises his structure in imagination before he erects it in reality."[15]

Frederick Engels in *Dialectics of Nature* provides a complementary argument. He maintains that technologies enable a person to "impress his stamp on nature,"[16] thereby changing the environment. Instead of emphasizing the planning in advance, Engels places importance on the future results of labor with technologies, as they are evidenced in the world. While Marx values the intentions at the production phase prior to the usage of technologies, Engels emphasizes the effects of that usage on the world.

More than one hundred years later, Bernard Stiegler offers a third perspective on tools according to which a tool is characterized by its "longevity." It means that a tool is an object that is preserved and passed on to the next generations. This observation focuses on the repetitive usage of the same artifact and its passage from one generation to another (obviously the same should apply to the underlying technology that is the basis for the specific tool-object). Stiegler emphasizes this point in order to show how history is created and how central the role of technologies is in this historical process.[17] Therefore, for Stiegler "A tool is, before anything else, memory."[18] Tools serve as durable memories that preserve historical modes of labor. What is missing in Stiegler's argument is the element of intention and planning which is dominant in Marx's analysis. Along this line, it is not clear if a tool qualifies as memory if (and only if) it is fabricated in order to preserve a memory or practice, or whether it is sufficient to use it as an intragenerational memory *ex post facto*.

So far we have seen that the question "what is a tool?" has been answered by Marx, Engels and Stiegler with an orientation toward the future. Another element in these answers is a concept of labor that is constitutive for humanity. Marx's analysis focuses on *Homo Faber* and sets for labor a central role in human life. In his analysis the tool is fundamental. He writes, "the first thing of which a labourer possesses himself is not the subject of labour but its instrument. Thus, Nature becomes one of the organs of his activity, one that he annexes to his own bodily organs."[19] The tool, according to Marx, is embodied to such an extent that the worker experiences the tool as an organ of his or her body. To use Donna Haraway's term, Marx's laborer becomes a "cyborg," even in relation to the simplest early tools.[20] Unlike Marx who regards the tool as part of one's body, Heidegger coins the term readiness-to-hand in order to refer to the "normal" and automatic usage of a tool in which the tool withdraws to the background and becomes part of it. He terms that background as Worldliness (*Umwelt*). The tool becomes a node in a network of objects that form the background world and is the context in which one operates. Marx, in contrast, places the tool in the foreground as a part of a hybrid that composes the human laborer and her tool. It should be noted that the concepts of the hybrid and the network are not unique to tools, and have also been used to characterize machines and even DTs.

Whether the tool is a part of the body or that which withdraws to the background, in both cases the hand has been conceived as the basic means of work since Marx and Engels. The hand is not only the primary means of work; Engels maintains that since prehistory "the specialization of the hand . . . implies the tool, and the tool implies specific human activity; the transforming reaction of man on nature, production."[21] Engels identifies an evolutionary process between the hand and the tool that produces a change in the world. The hand in this process is both the interface with the tool and its source. As an interface, the hand is what differentiates the tool (the *what*) from its bearer (the *who*). Hence, it becomes clear why Heidegger uses the word "hand" in his notions of readiness-to-*hand* and presence-at-*hand*. These notions do not only refer to the background and foreground with respect to the user, they also involve a hand that attaches the tool to the human. Jacques Derrida in his essay "Heidegger's Hand" finds an additional and important role for the hand in Heidegger's thought. The hand for Heidegger, according to Derrida, is what makes thinking possible.

But the hand alone cannot operate a tool. Engels recognizes that the hand needs to be matched with the brain. He describes the hand and the brain as coevolving organs: "step by step with the development of the hand went that of the brain."[22] The two are joined because "the hand alone would never have achieved the steam engine if, along and in parallel to the hand, and partly owing to it, the brain of man had not correspondingly developed."[23] The tool, the hand and the brain form a trio in which each part co-shapes the others and is codependent on them.

Paleontologist-anthropologist Andre Leroi-Gourhan, who studied prehistorical humans, adds the skeleton to the trio of tool-hand-brain. The hand

and the brain, according to Leroi-Gourhan, are not sufficient to describe the historical development of tools. It is the upright posture that frees the hand and changes the point of view. That is why Leroi-Gourhan asserts that "Tools and skeletons evolved synchronously."[24] While the skeleton and the brain play important parts in the codevelopment of humans and tools, they do not directly participate in the majority of usage modes. Yet, they are both of prime importance to tool analysis (whether we discuss tools, machines or DTs) because adding them leads one to consider the body as a whole. The use of a tool (with the hand) yields not only a change in the hand (through the development of skills) but also determines how one carries one's body; likewise, one's body position bears upon how a tool is used.[25]

From Tools to Machines

The *Oxford Dictionary* broadly defines a machine as "an apparatus using mechanical power and having several parts, each with a definite function and together performing a particular task."[26] This definition covers not only modern machines like trains and sewing machines but also medieval clocks and even ancient mills which were mechanically operated by wind or water. If we follow this broad definition, machines cannot solely be classified as "modern" and hence we cannot determine that tools are exclusively "premodern." Even Marx in *Capital* notes that the difference between tools and machines cannot be identified "by hard and fast lines of demarcation."[27] Just like machines existed in premodernity, so too do tools exist in modernity; "the tool . . . is not exterminated by the machine."[28] Both the tool and the machine have existed in parallel. Yet, the machine is different from the tool and is "emblematic of Modern Industry."[29]

Inspired by Charles Babbage's ideas of computers, Marx compares the nineteenth century's machine to the human body and identifies three sets of resemblance: the motor corresponding to the heart; the control and transmission parallel with the head or the brain; and the working parts equivalent to the hands.[30] Marx emphasizes the last component because these tools-hands are "the machinery with which the industrial revolution of the eighteenth century started."[31] He defines the working parts as "a mechanism that after being set in motion performs with its tools the same operations as the worker formerly did with similar tools."[32] The machine replaces not only the tool but also the worker. It is an anthropomorphic entity that uses its own tools. Marx writes, "capital now sets the labourer to work, not with a manual tool, but with a machine which itself handles the tools."[33] Tools, even when they are compared with human organs, do not gain the amplified anthropomorphism of the machine. Hence, the machine "becomes a competitor of the workman himself."[34] While the tool is considered by Marx to be an extension or prosthesis of the worker's cyborg body, the machine is deemed by him to be a rival to be fought against.[35] The relationships between humans and machines are not harmonious but hostile.

The sources of this hostility can be studied from the evolution of the relationships between the machine and the hand as outlined in chapter 15 of *Capital*. At the beginning, handicraft gave way to "manufacture." Donald MacKenzie notes that Marx uses the term manufacture in its literal sense: *manu*—hand in Latin; *factura*—to work or to act. At this stage, the machine substituted muscular power, to use Marx's term, and the hand was necessary for the machine's operation. Marx equates this type of machine with the worker; both are tool performers. In the second phase, as machines evolved, the manufacturing process was revised and the factory emerged. In this modern stage, machines in the factory substituted the hand, and the hand remained necessary only for the construction and maintenance of the machines. Once the human laborer is replaced, machine's productiveness is "measured by the human labour-power it saves."[36] The substitution then, is not only of the hand but of the worker as a whole. This is how the modern factory transformed the relations between workers and the instruments of labor, according to Marx. This understanding is the essence of his critique of capitalism.

From a technological perspective, the difference between tools and factory machines can be defined by the characteristics of the energy necessary for their operation. Tools are operated by "inconstant and uncontrollable" sources of energy,[37] be it a hand, an animal, wind or water; by contrast, machines are dependent on power that is "entirely under man's control,"[38] working in accordance with the laws of thermodynamics (electricity is an important development of the age of machines, but its emergence does not amount to a paradigm shift). Thanks to new sources of energy, the machine is independent of natural forces and aims at controlling and subordinating nature to a new order.[39]

Another difference between tools and machines, according to Marx, is rooted in the invisibility and imperceptibility of the engine and the transmission. The tool is mostly visible to the worker, who is likely to produce his or her own tools and to think of how to improve them; the machine's mechanisms—even if open and observable—require more than a layperson's know-how. The average worker cannot think of how to improve or invent new machines because the knowledge required for the production of a machine is entirely different from that which is required for its operation.[40]

Marx further maintains that the machine operates on a different scale: first, it produces larger quantities as it can simultaneously work on several items. Second, the machine can operate larger tools like "a gigantic razor" or "a monster pair of scissors."[41] Third, the machine runs at higher speeds. Finally, the "extensity" of the machine is greater in comparison to the human worker. That is, the machine is more forceful, as it is "entirely emancipated from the restraints of human strength,"[42] and functions with better precision.[43] As a result, the machine's product is qualitatively and quantitatively different from the tool's product.

In conclusion, the distinction between machines and tools is attributed by Marx to the autonomy of the machine's energy from natural constraints, the know-how of the machine's mechanics and the machine's scale of work. But the most important distinction is one of means-end relations. Marx writes, "in handicrafts and manufacture, the workman makes use of a tool; in the factory, the machine makes use of him."[44] Whereas humans apply tools in accordance with human needs, when it comes to machines, humans attend to the machine's needs of power, raw materials and maintenance. In other words, while the relations between humans and tools are defined by human usage, the relations between humans and machines are determined by the machine's requirements, so that "the labourer becomes a mere appendage."[45] Bernard Stiegler describes this situation as "the human is no longer a 'tool bearer',"[46] hinting at Franklin's definition of the human as a tool-making animal, quoted by Marx. This is a reversal of means-end relations. No longer do tools serve humans; now humans service machines. The traditional understanding of tools is overturned by machines. Subject-object hierarchy is inverted as well as means-end order, and humans operate for the benefit of the machine.

From Machines to DTs

What I call DTs are the contemporary technologies described in Negroponte's, Castells's, and Ihde's writings. These technologies consist of hardware and software, and include artifacts such as computers, cell phones, and hybrids of them (e.g., tablet). Television, although historically located in the age of the machine, can also fit into the definition of DT. In Negroponte's terms, bits instead of atoms are moved; thermodynamics is replaced by electronics; and the output is nonphysical and informational.

DTs maintain new kinds of relations with their users, different from the relations laborers maintained with machines. The differences in the relations can be mapped into five interconnected aspects: the operating organs, the types of output, the investment required from the workers, the site of the relations, and the general control scheme.

The first aspect concerns the operating organs. While the hand is a dominant organ for tools and early machines, the fingers of the hand became the dominant organs for DTs. The fingers are needed to type on a keyboard, the thumb is necessary for texting,[47] and the index finger is required for selecting from a list or activating applications from a touch screen. Moreover, the hand that operates tools exercises some force and tends to engage the whole body; the fingers that operate DTs use minimal force and have minor effect on other body organs. As these technologies evolve, some DTs become voice-operated, making the fingers redundant (e.g., iPhone 4S's Siri or voice-activated remote control for television). At the same time, eyeball movement detection techniques are being developed, so that it may be possible to operate devices just by looking at them. What unites voice operation and eyeball tracking techniques is that the hands are marginalized and the face becomes central.[48]

Second, the shift in the operating organs is accompanied by a shift in the ways in which technologies respond to humans. It is a shift in the outputs of technologies that move from the auditory to the visual. The auditory output characterizes tools and machines, as can be evidenced, for example, by the noise of hammering a nail, or the scratch of a pen on a paper; for machines it can be evidenced by their own working sounds, like the humming of the engine in the factory, the ticking of the clock, or the roaring of the car (especially in diesel cars, not so much for electric cars)—these are the sounds of the working of tools and machines. In contrast, DTs operate silently. If and when they produce noises, these are planned and designed, such as radio broadcasts, TV programs and a cell phone's ringtone. Instead of the noise, what is dominant for these technologies is the screen. Even radios in contemporary cars visually display the various levels of voice and the name of the radio station being listened to; the cell phone has a screen as large as the palm of the hand. The silence of DTs and the increased importance of the visual is what I call "the victory of the visual over the auditory."[49]

Third, in the context of labor, the workers and the working environment for DTs have changed. If tools were operated in workshops and machines in factories, DTs can be found in offices. But the new name and the much better physical conditions should not hide the fact that the proletariat still invests their time and effort as participants in the wealth-production process.[50] Workers are still situated adjacent to means of production, though the means of production are not necessarily machines but more likely to be computers. The fundamental difference is that whereas the nineteenth-century laborer invested physical efforts in front of the machine, the high-tech employee invests his or her "mastery over nature" in front of DTs.[51] DTs require more than muscle work; they require the employee's intellect, attention and enthusiasm. This is what Bifo Berardi critically calls "cognitive labor." In his book *The Soul at Work* he claims that the social structure remains basically capitalistic and so the profits flow to the hands of the few, even though the nature of work has changed.

Fourth, the analysis of technologies is no longer limited to labor relations. Expanding beyond the workshops and the factories, researchers need to follow DTs to the homes and public spaces. Hence the analysis is performed in terms of users, consumers and viewers, and not only of workers acting with technologies. The home becomes a major meeting point between the human and the technological. The human does not need to go to the factory or to the big city in order to meet the technology; DTs feel at home in the domestic and the public spheres. They are everywhere.

Finally, DTs set up new relations between humans and their technologies. While tools were perceived by Marx to be in the control of humans, and humans to be in control of machines, Gilles Deleuze and Felix Guattari argue that humans and DTs are components of the same system. They learn from the domestic space how these new relations function:

> For example, one is subjected to TV insofar that one uses and consumes it, . . . ("you, dear television viewers, who make TV what it is . . ."); . . . But *one is enslaved by TV as a human machine* insofar as the television viewers are no longer consumers or users, nor even subjects who supposedly "make" it, but *intrinsic component pieces*, "input" and "output," feedback or recurrences that are no longer connected to the machine in such a way as to produce or use it.[52] (emphasis added)

According to Deleuze and Guattari, the links between humans and DTs are symmetrical—humans use and consume DTs while being used and consumed by DTs. It is no longer a viewer or a user who has a separate distinct stance from which she can be alienated or alienate. Likewise, television is not a means to an end, something in order to, as the tool was, nor is it an omnipotent entity that invents tools and attempts to control nature. The relationships between the television and the viewer transcend the usage or subjection patterns of the machine. Watching television requires very little operation or control. Instead, the viewer is required to "play" her part, along with the TV set. The human and the technological artifact operate in a complex system in which they are no more than components. Deleuze and Guattari put this new type in the context of what they call the third age of technical machines:[53]

> [I]t is the reinvention of a machine of which human beings are constituent parts, instead of subjected workers or users. If motorized machines constituted the second age of technical machines, cybernetic and informational machines form a third age that reconstructs a generalized regime of subjection . . . *the relation between human and machine is based on internal, mutual communication, and no longer on usage or action.*[54] (emphasis added)

More radical than modern machines, DTs are not necessarily at the service of the human. These technologies call for a redefinition of their relations to the human and it is difficult (if not impossible) to determine who and what is superior.

POSTPHENOMENOLOGY

Locating the Discipline

The cell phone is a prominent DT. Like other technologies, it is studied in various disciplines, including sociology, history and philosophy. Each discipline has its own methods and perspectives. Let us take sociology as an example. Sociological approaches attempt to be as accurate as possible and so many of its field studies tend to look at a certain segment. They limit themselves to certain localities, specific time spans and particular behaviors and customs.[55] This accuracy comes with a price; most of these studies hardly provide a general framework by which we can begin to understand the rich array of relations between humans and cell phones in other geographies, cultures, or ages. If we switch to historical approaches, it may seem that they solve the tension between details and the wide perspective. However, these approaches have a major limitation, which is inherent to its mission: historical research studies the past. Such research provides useful methods to examine a remote history rather than the near-past. For the cell phone research both disciplines provide rich data but as this DT rapidly evolves, the sociological and the historical research tend to quickly become outdated.[56]

Although philosophy has its own drawbacks, certain branches of it attempt to integrate the conclusions of sociological and historical studies, thereby providing a wide perspective of contemporary technologies. In this book I employ a philosophy of technology that is known as postphenomenology. Founded on phenomenology and pragmatism, this school of thought integrates field research of various disciplines—including sociology, archeology and many others—in order to contextualize its findings in the world. At the same time, postphenomenology is genealogical by its nature, as it examines the changes in the relations between humans, technologies and the world. Unlike historical research that tends to explore the linearity of evolutions, postphenomenology investigates the genealogical aspects, thereby revealing the "rhizomatic" structure of technological developments.

This postphenomenological study of the cell phone accumulates the findings of sociological and historical research as sources of data, together with handset manuals, journalistic articles, blog posts and the like. From this mixture I extract insights that are relevant for the present and the future and that are free from the imposition of linearity. Postphenomenology enables "capturing" several aspects of the multiplicities that cell phone research reveals and provides conceptual frameworks for their analysis.

Multistability

When approaching a complex technology like the cell phone, it is important to analyze the multiplicity of sub-technologies, of their usages and of their effects on the world. Postphenomenology endeavors to model such multiplicities. One of the major concepts developed for this purpose is multistability. It indicates that the same technology can be used differently by different people who assign different meanings to it. The varied usages can be dramatically different from the planned usage of the technology's original design. Furthermore, even if people share the same purpose, they might use different technologies to pursue that purpose. Thus, although the cell phone was originally developed to make calls while away from a landline telephone, it has also been used for communicating by text, as well as for the digital writing of short texts like contact names, memos, to-do's, calendar entries, etc.

Multistability denotes that numerous purposes and applications can be ascribed to a single technology, be it a tool, a machine or a DT. The concept of multistability is applicable to every technology because each has "structured ambiguities that allow what first appears as a 'same' technology to be differently situated and have different trajectories."[57] Different people assign different meanings to the same technological artifact because their cultural backgrounds differ. It means that the same technological feature can serve different purposes. The cell phone is a good example for the different trajectories and cultures that enable diverse usages: children (and some adults) regard the cell phone as a gaming machine; teenagers and adults as a communication device; and parents as a controlling apparatus over their offspring from afar. Even a simple function such as texting has multistability, as it can serve not only to communicate with others but also to report the weather forecast or traffic alerts; it can even be used as a self-reminder tool.

When Heidegger posits that a tool is "something in order to," he ascribes a single function possible for it. The hammer for him is only for hammering. He does not ascribe multistability to such a tool. However, the hammer can have different usages in other contexts, which produce different meanings: pinning nails in the wall, mending shoes, a sculpting tool, a paper weight and even a murder weapon.[58]

In the context of multistability, tools and DTs significantly differ quantitatively. Ihde argues that contemporary technologies have "drastic multistability" embedded into their very existence.[59] Whereas we need to exercise some creativity to invent more usages for the hammer, DTs enable many usages from the beginning. Ihde's example is the computer's visual display screen that is used for presenting a text typed with a word processor, a virtual world of a computer game, or a video. Likewise, the cell phone, which was originally designed to facilitate voice communications, is also used to store our address book, wake us up, remind us of meetings and events, and more. The cell phone's multistability produces and is produced by the different relations it maintains with its users.

Human-Technology-World Relations

The multiplicity of sub-technologies, their usages, and their effects is modeled by the postphenomenological scheme of I-technology-world. This scheme focuses on the relations between its three components, thereby opening a unique perspective that is different from those produced by other disciplines such as sociology and history. From this scheme, Ihde identifies and maps four relations:

- **Embodiment relations**, in which we experience the world via a technology that becomes part of us, for example, the eyeglasses through which we see. Similarly, the cell phone becomes part of us, merging with our ear and mouth, when we talk with a friend;
- **Hermeneutic relations**, in which the technologies used require a "reading" of the world and the user's attention is on the artifact rather than on the "real world," a thermometer, for example. Likewise, the GPS function of the cell phone indicates our location;
- **Alterity relations**, in which technology stands as an object next to us, acting as an avatar, for example, an ATM, or an automatic answering service for a cell phone; and
- **Background relations**, in which the technology operates in the background almost unnoticed, for example, the humming of a refrigerator, or the fan of an air conditioner.

The multiplicity of relations encompasses the contradicting approaches of Marx and Heidegger, where the former views technologies as part of the user's body while the latter conceives them as part of the background. The various relations demonstrate how the same technology sometimes maintains embodiment relations in which the technology becomes part of the user's body and sometimes maintains background relations and withdraws to the background.[60]

Historical Variations

Another dimension of multiplicity in the analysis of technologies is diachronic by nature; it attempts to model the stages of evolution of the technology, as well as of the other components of the I-technology-world scheme.

Postphenomenology's notion of historical variation assists in the conceptualization of the varied relations between people, technologies and their environments as they evolve over time. Although termed historical variations, the variations are closer in their meaning to genealogy than to history because they examine the relations between humans and technologies by focusing on the "how" instead of the "what."[61] They do so by grouping historical and sociological data into genealogical periods, thereby characterizing the distinctiveness of the relations unique to each period. In Ihde's words, historical variations "divide [data into] periods by the style of technologies used" while simultaneously reviewing the "practices that correlate with the technologies."[62] Each historical variation is differentiated by technological functionalities, their related practices and their effects in the world. Consequently, each historical variation is distinguished by a certain "Zeitgeist," or set of worldviews. Neil Postman beautifully describes this phenomenon with regard to tools and DTs: "to a man with a hammer, everything looks like a nail . . . To a man with a computer, everything looks like data."[63] Here Postman and Marx meet, as they both show how the employment of technologies shapes our *worldview*. Postphenomenology structures this shaping into changes in our *world* and changes in our *views* (and practices).

Marx's tool analysis demonstrates how certain technologies, views, practices and environments form historical periods, what I call here historical variations. Specifically, Marx analyzes the technologies, the work practices and the work environments of ancient times, the feudal period and of modernity. Heidegger follows this type of analysis when he contrasts the Classical Greek period represented by a silver chalice with modern times represented by a hydroelectric plant on the river Rhine. A similar classification is presented by Ihde in his analysis of astronomical technologies. He characterizes ancient astronomy by practices based on the bare-eye gaze, early modernity by Galileo's invention and use of the optical telescope, and the contemporary period by the use of an infrared telescope combined with imaging technology. In my research on the cell phone I pursue this type of analysis with a modification. Instead of considering several technologies that are representative of different technological ages, I focus on a single technology—the cell phone—and divide it in to historical variations that each lasts no more than several years. Hence my variational analysis in this book includes some prominent cell phone models—Motorola's StarTAC, Nokia's 5110, and Apple's iPhone. Each represents a historical variation of cell phone technology.

Invariants

Once the variational analysis is complete, the next phase in the postphenomenological investigation is the "extraction" of invariants.[64] Invariants can be regarded as the common denominators of variations. Unlike variations that constantly change, invariants remain fixed. Invariants are "structural patterns."[65] They are like the warps that compose the carpet's foundation yet remain hidden underneath the beautiful patterns of the wefts. Invariants function as "correlational rules" that pass between the historical variations. They are hidden for most users, but can be revealed by postphenomenological examinations.

The notion of invariants provides explanations for cell phone phenomena as well as for the question of why a cell phone remains a cell phone even though the technology, its modes of use and the world around it have significantly changed. It also assists in answering the question of why this technology is indispensable for many of its users.

The notion of invariants should not be conflated with the Heideggerian essence. Indeed, they both tend to be hidden, but invariants are plural while an essence is one and only one. For Heidegger the singular form is not a grammatical imposition but instead a well-selected wording conveying the message that technology can have no more than one essence. What I seek to show in this book are the multiple aspects of the cell phone. I attempt to prove that the essence of a cell phone is not limited to that of a telephone without a cord. The invariants will assist in showing the multiplicity of experiences that the cell phone mediates for us. As "structural patterns" or warp threads, there can be many, but not infinite, invariants. They assist in producing additional perspectives of the technologies investigated.

Histories of the Present, Histories of the Future

As a genealogy-oriented theory, postphenomenology provides an important analytical arsenal for the examination of the relations between cell phones, users and the environment. These tools include: the notion of multistability that frames the multiplicity of usages and meanings for a single technological artifact; embodiment, hermeneutic, alterity and background relations that describe different usage schemes; and the historical variations as well as the invariants that help in identifying the different patterns of using cell phones. This postphenomenological arsenal is suitable for the study of tools, machines and DTs. DTs, however, require an additional concept that refers to their acceleration toward the future, which shapes a specific present.

Michel Foucault speaks of the "history of the present" to emphasize the present perspective of previous historical periods. According to Foucault, genealogies of the past change in accordance with the current point of view. Genealogical research therefore maps historical variations in order to understand a current experience of the world.

This postphenomenological analysis of the cell phone needs to switch from present to future because the pace of transformation for DTs is so fast that by the time a genealogy is fully mapped, the present point of view is already outdated. The current pace of change with regard to technologies makes it impossible for researchers to map a genealogy that remains meaningful. To use Derrida's term, "time is out of joint"[66] and so it is extremely difficult to compose a history of the present when it comes to mapping DTs' evolution.

Histories of the future can be understood within the Heideggerian temporality which ties the present with the future so that "the future . . . makes present in the process of having been."[67] It is the future that produces the present. The present is not merely positioned between a past and a future, but is a "comportment" toward the future. In this process, Ihde explains, "a projected possible future gathers a particular perspective upon the past into a then finally constituted present."[68] The past is interwoven into the future as a source of perspectives, so that our understanding of the past reflects our present position and the "comportment" toward the future. While attempting to describe and qualify such "a projected possible future" for the cell phone, this book is not engaged in "technological dreams" or "technomyths."[69] Instead, it constitutes a search for the roots of these future possibilities in the very recent past, while refraining from science fiction and imaginary scenarios.

Both histories of the present and of the future call for an inquiry into the conditions for experience. But instead of studying the actual technologies used in the present or contemporary scientific knowledge, the focus should be on the potentialities that are waiting to be realized, referring to the present as a condition for the future. The starting point of the analysis of histories of the future is the practical which is present and unstable, what philosophers would term contingent. The target is also different: histories of the future do not seek to discover structures of power through discourse; instead they remain in the domain of everydayness and at the level of experiences. This is not necessarily a political project in the strict sense of the word but an effort to understand how we experience the world (see Table 1.1).

Table 1.1 Histories of the Present, Histories of the Future

History of the Present	*Histories of the Future*
Explores the past as a condition of the present	Explores the present as a condition of the future
Looks for facts and documents	Looks for potentialities and trajectories
Studies the sciences	Studies technoscience
Focuses on knowledge production	Focuses on technological evolution
Genealogy of values	Postphenomenology of sensations
Analysis of discourse and power	Analysis of embodiment and experience

THE PLAN OF THE BOOK

Historical variations and invariants form a matrix that can be expanded in all directions. There will always be future historical variations, and additional past variations can be identified. More invariants exist; they just wait to be revealed. And those which have already been identified can be reordered and reworked. For reasons of clarity and simplicity, the book is divided into two main parts: the first deals with the historical variations and the second with the invariants.

The first part provides an overview of the cell phone's evolution through four historical variations. This part is, in a sense, a Foucauldian project that processes nonlinear historic data and assembles it into histories of the present and of the future. The first historical variation is titled "Talking Heads" and it concerns voice. In this variation the cell phone was designed for voice communication. Early texts used in these cell phones will serve to show the history of the future of the next variation which is described in the chapter "Texting-at-Hand." This second historical variation is focused on the cell phone's textualities. Unlike most of the sociological and historical research on the cell phone that focuses on texting (i.e., communication via short text messages), this chapter examines also other textualities such as memos and calendar entries. The following chapter, "The Kingdom of Multimedia Applications," is a description of the third historical variation which is centered on multimedia applications. The next chapter, titled "Sensory Exploration," examines the technologies, meanings and practices of augmented reality and how the cell phone assists in sensing the environment.

The second part of the book contains three of the cell phone's invariants, each serving as an insight that illuminates the variations. Each chapter presents an invariant. The first invariant is the wall-window. It presents how cell phones of all historical variations provide an imaginary wall for their users against the immediate environment, partially separating the user from her surroundings. At the same time, the cell phone opens a window to a remote space, whether it is real or virtual. The next invariant is the quasi-face. This chapter aims at explaining why the cell phone has a screen through an analysis of Levinas' notions of the Other and the face coupled with the postphenomenological notion of alterity relations. The argument in a nut shell is that if technology is a quasi-other, then the screen can serve as a quasi-face. The third invariant discussed is the memory prosthesis. Based on the work of Stiegler who refers to tools as a prosthesis and as a memory recorder, both terms will be combined and analyzed.

Ihde usually refers to the embodiment element of experience as an invariant; the three invariants addressed in this book are related to the other three postphenomenological relations: The hermeneutic relations are the condition of the wall-window invariant, because the cell phone has to be "read" in order to create a window. Alterity relations manifest through the quasi-face invariant, representing a quality of the cell phone that enables us to refer to the cell phone as a quasi-other. Lastly, background relations can be traced in the memory prosthesis invariant, in which the cell phone becomes part of the human memory albeit externalized. However, the link between a certain invariant and a postphenomenological relation is no more than a starting point. In the discussion of the various invariants related to cell phone models I touch upon questions of embodiment and mix together these relations in order to create a rich picture for each invariant.

The last chapter attempts to form one possible invariant of invariants termed "becoming-mobile," which is based on Deleuze and Guattari's notion of becoming. Becoming-mobile is a conceptualization of mobility with or without movement. It encompasses situations where physical movement does not exist, such as the movement from one space to another as performed in phone calls, when watching a video, or when playing a game. Becoming-mobile models new forms of mobility, like those of augmented reality, where moving atoms and moving bits maintain transductive patterns in which one influences the other.

NOTES

1. Nicholas Negroponte, *Being Digital* (New York: Vintage Books, 1996).
2. Manuel Castells, *The Rise of the Network Society* (Malden, Oxford, Victoria: Blackwell Publishing [1996] 2000).
3. Castells, *The Rise,* 1.
4. Ibid., 3.
5. Eight years after publishing this book, Castells and other scholars published research on the social change brought by cellular technology (Castells, Fernandez-Ardevol, et al., *The Mobile Communication Society: A Cross-cultural Analysis of Available Evidence on the Social uses of Wireless Communication Technology,* A research report prepared for the International Workshop on Wireless Communication Policies and Prospects (2004); published as a book in 2007). Not only is the Internet a central component in the Information Age, but also the cell phone.
6. Don Ihde, *Postphenomenology: Essays in Postmodern Context* (Evanston: Northwestern University Press, 1995).
7. Don Ihde, *Heidegger's Technologies: Postphenomenological Perspectives* (New York: Fordham University Press, 2010), 5.
8. Other philosophers have also argued that contemporary technologies are dramatically different from traditional and modern technologies, for example, Friedrich Jameson, Jean-Francois Lyotard, Bernard Stiegler, Paul Virilio, Zygmunt Bauman, to name a few.

9. Don Ihde, *Expanding Hermeneutics: Visualism in Science* (Evanston: Northwestern University Press, 1998), 178.

10. Karl Marx, *Capital: A Critique of Political Economy* (International Publishers, 1983), 179.

11. Marx, *Capital*, 179.

12. Martin Heidegger, *Being and Time* (Albany: State University of New York Press, 1996), 64 [68].

13. Heidegger's view that a tool has a clear and concrete goal is criticized by Ihde. Ihde stresses the inability to limit a tool to one purpose and introduces instead the notion of multistability—see the detailed discussion in the Postphenomenology section of this chapter.

14. Marx, *Capital*, 179.

15. Ibid., 178. Cf. Donald MacKenzie, "Marx and the Machine," in *Knowing Machines: Essays on Technical Change*, ed. Donald MacKenzie (Cambridge and London: The MIT Press, 1996), 23–48.

16. Fredrick Engels "Dialectics of Nature," in *Karl Marx and Fredrick Engels Collected Works (Vol 25) – Friedrich Engels: Anti Duhring and Dialectics of Nature* (International Publishers, 1987), 330.

17. A similar explanation is provided by Michel Callon and Bruno Latour, who suggest that humans, unlike baboons, attach social meanings to objects, which they name "a Leviathan." Only a human society can have a Hobbesian Leviathan, that is—a macro-social actor composed of multiple micro-social actors that are durable across space and time. In the context of tools, an object becomes a tool if it is important enough to be passed to future generations. Such importance is evident from the fact that it carries a meaning to the members of society.

18. Bernard Stiegler, *Technics and Time, 1: The Fault of Epimetheus* (Stanford: Stanford University Press, 1998), 254.

19. Marx, *Capital*, 179.

20. Donna Haraway, "A Cyborg Manifesto: Science, Technology, and Socialist-Feminism in the Late Twentieth Century," in *Simians, Cyborgs and Women: The Reinvention of Nature* (New York: Routledge and Institute for Social Research and Education, 1991), 149–181.

21. Engels, "Dialectics of Nature," 330.

22. Ibid.

23. Ibid.

24. Quoted by Stiegler in *Technics and Time, 1*, 154.

25. Cf. Don Ihde, *Technology and the Lifeworld: From Garden to Earth* (Bloomington and Indianapolis: Indiana University Press, 1990). This is the meaning of postphenomenological embodiment relations (to be detailed later in this chapter). This understanding is also the foundation of Deleuze and Guattari's notion of "Body without Organs." It implies that a body is more than an aggregation of biological organs or organs used with a particular tool. This is in line with postphenomenology's understanding of the body as whole, an assemblage that exceeds the aggregate. For example, when Galileo looked through his telescope, seeing involved not only the eye that looks through the tool, and not only the hand that holds it. The hand-brain,

supplemented by the eye in this case, and "fused" with the telescope, creates a new embodiment, a body that is beyond organs and yet can sense and understand.

26. http://oxforddictionaries.com/definition/english/machine?q=machine (accessed 8 January 2013).

27. Marx, *Capital*, 371.

28. Ibid., 387.

29. Amy E. Wendling, *Karl Marx on Technology and Alienation* (Basingstoke: Palgrave Macmillan, 2011), 137.

30. Marx, *Capital*, 373; cf. Wendling, *Karl Marx*, 138.

31. Marx, *Capital*, 373.

32. Ibid., 374.

33. Ibid., 387.

34. Ibid., 430.

35. Ibid., 427. In this war, the machines always win. It is like "an encounter between an army furnished with breach-loaders and one armed with bows and arrows" (450). However, Marx notes that when machinery is considered alone it can be regarded as producing positive outcomes, such as shortening hours of labor and lightening workloads. When the machine is in the service of capital, it produces opposite results—lengthening the working hours, heightening the intensity of labor, etc. Hence, the context of use should lead the analysis of technological artifacts.

36. Marx, *Capital*, 391.

37. Ibid., 377.

38. Ibid., 374.

39. Cf. Martin Heidegger, "The Question Concerning Technology," in *The Question Concerning Technology, and Other Essays*, trans. W. Lovitt (New York: Harper and Row, 1977), 3–35. For Heidegger machines require energies that can be not only controlled but also accumulated "in reserve."

40. See Marx, *Capital*, 486; MacKenzie, "Marx and the Machine," 28.

41. Marx, *Capital*, 385.

42. Ibid., 378.

43. See also Wendling, *Karl Marx*, 142. For example, Charles Babbage owned a woven portrait of Joseph Marie Jacquard made of one thousand threads per inch, produced with 24,000 punch cards; see Sadie Plant, *Zeros + Ones: Digital Women + The New Technoculture* (New York: Doubleday, 1997), 16. Such delicacy and precision can hardly be accomplished by tools.

44. Marx, *Capital*, 422.

45. Ibid., 386.

46. Stiegler, *Technics and Time, 1*, 69.

47. For example, "The Thumb Generation," coined by Sadie Plant, *On the Mobile: The Effects of Mobile Telephones on Social and Individual Life* (Motorola, 2003) or "Thumb Culture," coined by Glotz, Peter, Stefan Bertschi, and Chris Locke, *Thumb Culture: The Meaning of Mobile Phone for Society* (Bielfeld: Transcript Verlag, 2005).

48. For media technologies, from the book to the television, the hand has always been of secondary importance. What mattered most, and still matters, has been the face—especially as an aggregation of hearing and seeing organs. Media technologies,

I assert, mark the transition from hand-based technologies like the hammer, to face-based technologies like the Internet (see the chapter on the quasi-face).

49. Galit Wellner, "Wall-Window-Screen: How the Cell Phone Mediates a World-view for Us," *Humanities and Technology Review*, 2011: 77–104.

50. Nick Dyer-Witheford, *Cyber-Marx: Cycles and Circuits of Struggle in High-Technology Capitalism* (Urbana and Chicago: University of Illinois Press, 1999); Franco "Bifo" Berardi, *The Soul at Work: From Alienatio to Autonomy,* trans. Francesca Cadel and Giuseppina Mecchia (Los Angeles: Semiotext(e), 2009).

51. Karl Marx, *Grundisse: Foundations of the Critique of Political Economy* (London: Penguin, 1973), 705.

52. Gille Deleuze and Felix Guattari, *A Thousand Plateaus: Capitalism and Schizophrenia*, trans. Brian Massumi (Minneapolis, London: University of Minnesota Press, 1987), 458.

53. The first age is equivalent to the age of the tool, with the difference being that Deleuze and Guattari describe the relations as enslavement. The second age is equivalent to that of the machine and is described as subjection, because the human operator is subjected to the technical machine. In their works, the term "machine" refers to two meanings—one is a technical machine and the other is an apparatus, such as the state. Here, the first meaning is used.

54. Deleuze and Guattari, *A Thousand Plateaus*, 458.

55. For example, two high school students, eight college students, two housewives and three professionals, all residing in the Tokyo Kanto region (Daisuke Okabe, "Emergent Social Practices, Situations and Relations through Everyday Camera Phone Use," *Mobile Communication and Social Change* (Seoul, Korea, 2004)); 59 persons from 21 households (Richard Harper and Lynne Hamill, "Kids Will Be Kids: The Role of Mobiles in Teenage Life," *Mobile World* (2005): 61–74); 70 participants in ages 14 to 34 from Milan area (Barbara Scifo, "The Domestication of Camera-Phone and MMS Communication: The Early Experiences of Young Italians," in *A Sense of Place: The Global and the Local in Mobile Communication*, ed. K. Nyiri (Vienna: Passagen Verlag, 2003), 363–74); 88 students from School of Communications and Information Systems (Jamie Pinchot, Karen Paullet, and Daniel Rota, "How Mobile Technology is Changing Our Culture," *Journal of Information Systems Applied Research* 4, no. 1 (2011): 39); 93 persons living in a suburb of Oslo, Norway (Rich Ling and Leslie Haddon, "Machines that Become Us," in *Mobile Telephone, Mobility and the Coordination of Everyday Life*, ed. James E. Katz (New Brunswick: Transaction, 2003)).

56. Jason Farman, *Mobile Interface Theory: Embodied Space and Locative Media* (New York and London: Routledge, 2012).

57. Ihde, *Heidegger's Technologies*, 126.

58. Ihde, *Heidegger's Technologies.*

59. Ibid.

60. Peter-Paul Verbeek adds more layers of multiplicity as he identifies additional I-technology-world permutations. One is a hybrid intentionality in which the technology merges with the human, as in the case of an implanted microchip that enhances vision or antidepressants that change one's mood. Although the cell phone is not

physically integrated into the human body-like implants and pills, I will show how it is incorporated into the user's memory and intentions. The other relation that Verbeek forms is a composite intentionality which refers to the intentional dependency of humans and technologies. It can be demonstrated by augmented reality mobile applications that inform its user in advance of obstacles further down the road. Such technologies are dependent on the user's intentions and at the same time demonstrate human dependency for navigation instructions.

61. Michel Foucault, "Nietzsche, Genealogy, History," in *Language, Counter-memory, Practices: Selected Essays and Interviews*, ed. D.F. Bouchard (Ithaca: Cornell University Press, 1977), 139–164; Hubert Dreyfus, "Beyond Hermeneutics: Interpretation in Late Heidegger and Recent Foucault," in *Hermeneutics: Questions and Prospects*, ed. Gary Shapiro and Alan Sica (Amherst: University of Massachussets Press, 1984), 66–83; Colin Koopman, *Genealogy As Critique: Foucault and the Problems of Modernity* (Bloomington: Indiana University Press, 2013).

62. Don Ihde, *Postphenomenology and Technoscience: The Peking University Lectures* (New York: State University of New York Press, 2009), 46.

63. Neil Postman, *Technolopoly: The Surrender of Culture to Technology* (New York: Randon House, 1992), 14.

64. Don Ihde, *Experimental Phenomenology: Multistabilities (second edition)* (Albany: State University of New York Press, 2012), 46.

65. Ihde, *Experimental Phenomenology*, 22.

66. Jacques Derrida, *Specters of Marx: The State of the Debt, The Work of Mourning & the New International* (New York, Oxon: Routledge, 1994), 1.

67. Heidegger, *Being and Time*, 300 [326].

68. Ihde, *Expanding Hermeneutics,* 81.

69. Ibid., 105.

Part I

HISTORICAL VARIATIONS

Chapter 1

First Historical Variation

Talking Heads

MOTOROLA'S StarTAC 3000 (JANUARY 1996)

Commercial cellular communication services have been available since the early 1980s.[1] These services were operated through heavy handsets in a configuration known as the "brick" because of its size and shape. They were mostly used by business people, as can be seen in early pictures of cell phones populated with people wearing a suit and a tie holding a large brick handset. Because of their size, holding these handsets near the ear required some effort. By the 1990s, lighter and smaller handsets emerged in a form known as the "candybar" configuration. First, they were used by yuppies (i.e., young well-paid professionals) and then spread to almost everyone. Motorola, which was known for its brick handsets, introduced a small revolutionary cell phone in January 1996—the StarTAC 3000. It was the last major analog cell phone and was followed by digital versions in the years after. Although its electronic components were analog, digitality had inspired its design and functionality. This mixture of analog and digital is at the core of this chapter.

At the time of the initial launch of the StarTAC in 1996, its exterior design was regarded as unique. It was the first handset with a clamshell ("flip") configuration—a significantly visible differentiator compared to the brick or candybar configurations that were common at that time. The clamshell configuration produced new modes of use: opening the clam revealed the keypad and the screen, and also served to answer calls; closing the clam terminated a call; and when the handset was closed, the keypad was protected from accidental pressings. This configuration reminded *PC World*'s journalists of *Star Trek*.[2] Moreover, the proximity between the title of the movie and the model's name suggests that the designers of the phone at Motorola might have been inspired by that show.

With the clamshell configuration, StarTAC was small and lightweight, weighing just 3.1 ounces or 88 grams. Motorola was very proud of this feature and elected to open the "What's New" section in the User Manual with: "What would you say to a telephone that's smaller than your wallet? Hard to believe . . . if you hadn't already seen it with your own eyes." Thanks to its size and weight, StarTAC could be comfortably held in one's pocket or on one's belt. This embodiment also required it to be a robust, "pocket-proof" accessory that could survive—according to the first page of the user manual—even if users "cooked it, steamed it, shook it, shocked it, dusted it, and dropped it—and the phone still worked!" The placement of this statement and the exclamation mark were somewhat unusual compared to other manuals that hardly mentioned durability issues.

Another important function of the StarTAC was its ability to present the phone number of the calling party, known as "Caller ID." The ability to know who is calling before answering the call was considered novel. Yet, the service was subject to three conditions: (a) the network infrastructure supported Caller ID; (b) the user was subscribed to the service; and (c) at the handset level, the Caller ID functionality was turned on. Because these conditions required some technical know-how, they complicated the operation of Caller ID and made it inaccessible to many users. Today, however, knowing who is calling is much simpler to operate and this feature is considered elementary by most cell phone users.

Because of these unique features, *PC World Magazine* in its December 2005 issue mentioned the StarTAC in its "The 50 Greatest Gadgets of the Past 50 Years" list and ranked it in sixth place.[3] Other publications retrospectively crowned the StarTAC "as perhaps being the first truly modern mobile phone."[4] It remained up-to-date for a relatively long time, and Motorola revived it in 2004 as "Motorola StarTAC VIP 10th Anniversary Edition," keeping the same design and updating it with new materials and features (the sharp-eyed reader can notice that only eight years passed since the launch of StarTAC in 1996. . .). In an age when Moore's Law of eighteen months rules, an eight-year-old handset that keeps its appeal is not very common, to say the least.

THE USER'S PERSPECTIVE

Small enough to fit in the palm of the hand, "smart" enough to know who is calling, and wrapped in a novel clamshell configuration, StarTAC was embraced by users even before they purchased and utilized the phone:

"Nobody had ever seen a mobile phone like this before, and it was the first phone that you could easily put in your pocket or clip to a belt and not really notice it."[5] Although novel and exceptional in its time, StarTAC belongs to the first historical variation of cell phones and represents it because this device was—like most cell phones of its time—optimized for voice communication. Each of its distinctive features was oriented to facilitate the making and receiving phone calls. For example, the unique clamshell configuration turned out to be well suited for a "talking face" because it fit nicely between the mouth and the ear. When the cell phone was open, the microphone was close to the mouth—closer than in the candybar or brick layouts.

The number of calls received also increased thanks to the size of the handset. When the handset is large, it has to be stored in a bag or a briefcase, where ringtones are not always heard, and when heard it takes time to reach the handset. The small size and light weight made it easier to carry the StarTAC in one's pocket or on a "holster." The holster is an accessory that enables hanging the handset on the belt for a quick access. With the holster, the cell phone is embodied like a garment, which enabled Motorola to market it as a "wearable phone." The concept of the wearable phone entailed a new approach to the operation of the cell phone, as well as to the embodiment relations users maintained with it.

As a wearable phone, sound indication for incoming calls might be insufficient in noisy surroundings, so the auditory cue was supplemented by vibration. When the cell phone is held near the body, as the holster dictates, or when the cell phone is in the pocket, the vibration may be more effective than aural or visible indications. Vibration offered a new dimension for the embodiment relation between the cell phone and the user. It turned the cell phone into a multisensory device, not only heard or seen but also felt.

Another unique feature of the StarTAC was the introduction of a second battery. This allowed users to talk more while being away, and enabled more conversations with the cell phone.

The same goal led to the development of the Phone Book feature. Once a phone number is stored, a call can be made more easily, with no need to type the digits. But StarTAC's Phone Book required the user to remember the phone numbers, because it could not store or present names. In fact, the presentation of letters on the StarTAC's screen was not simple at all. Early StarTAC models had a small LED screen composed of tiny red lights that were lit in a certain composition to present digits. The screen could display no more than ten digits at a time, even if the number dialed was longer than ten. But the real limitation was on the display of letters. When preprogrammed text announcements appeared, the alphabetic characters were not of equal size or form. For example, the letters "m" and "w" required the space of two digits (see Figure 1.1); the letters "t" and "o" were presented in lower case;

Figure 1.1 StarTAC's Texts. *Source*: Author.

and "a" and "l" were in upper case. Consequently, the word "total" appeared as "totAL," and the word new as "nEw." Other words were shortened to fit the ten-digit display, so the indication "no messages" was compressed to "noME595," where the letter S was represented by the digit "5" and the letter G by the digit "9." With such a presentation of letters, it becomes evident why it was difficult to present the caller's name on that LED screen, and why the Phone Book stored only phone numbers without the associated names. The Phone Book contained up to 20 numbers, and the user had to recognize the numbers and do the hermeneutical work of associating each of them to a name.

StarTAC also included a Scratchpad feature. According to Motorola, it "lets you enter a phone number during a conversation: If you're on the phone and your party gives you another number to call, simply enter it into the Scratchpad to place the call later." Again, only numbers could be entered, the name and context had to be memorized by the user. Although it was intended to complement the Phone Book, this relatively free-form writing could have been useful for other purposes, such as memorizing other numbers like identification codes, credit card numbers, and numbered passwords.

The only way to display free-form text was in the form of a short message delivered as part of the nascent texting service.[6] This feature was inspired by pager service, which was well known at the time; StarTAC could receive short messages, but like a pager, could not answer them. It meant that a text message could be received only if the other party was not using StarTAC but some other cell phone model that supported messaging! This lacuna can be explained by the dominant conception of the cell phone as a voice machine, and it demonstrates (again) the supremacy of voice over text in the first historical variation of the cell phone. StarTAC was clearly voice-centric, and this lack of ability to write text messages is well reflected in its (in)capabilities.

The focus on voice is not only a functional design decision grounded in some historical-cultural background, but also has deep technical foundations, rooted in the architecture of the cellular networks in which the StarTAC operated. StarTAC's first versions were analog (the standard was AMPS, retrospectively categorized by the industry as "1G" or "first generation"). Experts observed that "It is interesting to note that Motorola chose to commit considerable effort to miniaturize an analog cellular phone . . . [while] outside the US the cellular infrastructure is rapidly adopting digital protocols."[7] Analog networks were optimized to carry voice, while digital infrastructure could easily carry both voice and text. Put differently, the underlying standards of StarTAC were voice oriented and support for non-voice services required significant research, development and design efforts. This may explain, among other factors, StarTAC developers' focus on voice.

THE IMPACTS OF THE FIRST HISTORICAL VARIATION

How did the cell phone of the first historical variation make a difference? To answer this question, let us follow the postphenomenological triad of I-technology-world and examine how each tier of the triad is transformed. In other words, the examination will pursue three aspects of transformation: how the sense of self is altered; how technologies evolve; and how the environment changes. It is important to note that within the structure of I-technology-world, all three parts of the triad are modified, albeit at different times, manners and speeds.

Basically, the cell phone's mobility mediates a world for its users and their interlocutors where calls can be initiated and taken everywhere (provided there is cellular coverage, of course). But the first historical variation's voice paradigm is more than a phone with mobility which enables us to talk while away from our home or office. At this stage, the cell phone has already transformed our reference to these very homes and offices. The distinction between dwelling and workplace is melting down, and it is no longer clear when I'm "at work" and when "with friends/family." For example, when my cell phone rings in the evening while I'm home, and I realize the call is coming from a colleague (in a different time zone? My boss?), is it "work time" or "leisure time"? In the morning my cell phone rings again. This time I'm in the office, working. The caller is my friend who wants to update me on her last shopping round. Is it "work time" or "leisure time"?[8] The cell phone is not alone in this melting-down process. The ubiquity of e-mails has a similar effect. Yet, the cell phone is one of the major contributors to this meltdown.

Not only are the boundaries between work and leisure eroded with the cell phone, but also the boundaries between being alone and being together.

The ability to walk (or drive) and talk with a remote interlocutor or to be alone in a hotel room and talk to family members produces what I shall call a "never-alone effect," where we can rarely feel alone.[9] Children of working mothers are no longer totally alone at home, as their mothers start using first-generation cell phones to wake them up in the morning, and later after school to make sure they have eaten lunch and done their homework. This is known in sociological terms as "remote mothering."[10] The never-alone effect can increase self-confidence. For example, when I walk in a dark street while talking on the cell phone, the voice of the interlocutor (as well as the glowing light coming from the screen) makes me feel safer. This sense of security can be also imputed to the fact that our friends and relatives have never been so close, requiring only the pressing of some buttons to hear their voices.

From a technological perspective, the cell phone moved away from the concept of "phone with mobility." In which direction? That was not very clear. Caller ID had a special role in the movement from the first to the second historical variation of cell phones, because it made a significant contribution (among other factors) to the evolution of the screen and memory. Here, in the first variation, screens and memory had modest roles. They enabled the elementary task of displaying the interlocutor's phone number and storing it for later use. Caller ID applications (such as Phone Book and Scratch Pad) required a cell phone that had a screen on which the identified number could be displayed as well as a small internal memory to store it. At this point, a text detailing the name of the caller was almost inevitable, as was the gradual increase in the number of contacts in the Phone Book. Consequently, a larger screen and a larger memory were required.

The growth in screen size and quality can be also explained by the embodiment relations users developed with the cell phone. With the screen, the cell phone's embodiment was no longer only auditory like a landline telephone, but also visual. The exclusivity of the auditory was also diminished by the vibration feature that added a supplementary sensory aspect to the cell phone's output. The cell phone became a multisensory technology; however, among the three senses of hearing, sight and touch, the visual eventually took over. Although cell phones of the first variation were not intended to be looked at, they were potentially gazed at for a moment to see a number in an effort to recognize who was calling, and then moved near the ear to conduct the phone conversation. This moment of gazing at the screen can be considered a baby-step toward a new type of interface between the cell phone and the user. The cell phone would become, what can be termed postphenomenologically, a quasi-other. The cell phone is a technological other with whom the user maintains alterity relations. It is a technological other in the sense that the user develops a complex relationship with the cell phone that is neither a relationship to a tool nor to a human. The relationship with the cell phone would become one of dependency, or better—companionship. As a technological other, the user looks at it. This look has led to a growth in the size and quality of the cell phone's screen from LED digits up to colorful displays.

As a quasi-other, the cell phone becomes distinct in that each handset is "personalized." In the first historical variation it is the address book that turns the cell phone into a personal artifact that cannot be handed to strangers without revealing private data such as which phone number I consider "home," who my partner is, or the significance of some friends (i.e., who is included in the 20 entries of the address book). Other aspects of personalization refer to the external parts of the handset, such as stickers of children or a beloved dog. Leopoldina Fortunati describes the evolution in the conceptualization of the cell phone from a mere mobile telephone to a personal artifact. She writes, "Despite the fact that the mobile phone was designed as a technology to be used while moving from one place to another, users have basically redesigned it as an instrument of individual communication."[11] She classifies the transformation of the cell phone from "mobile" to "personal" as a construct of the users, who shape and change the technological artifact by their acts. Perhaps a name like "personal phone," similar to "personal computer," would have been more appropriate for the cell phone. In the next historical variation the personalization intensifies and becomes more significant. In this historical variation, mobility still reigns.

As a personalized device that evokes quasi-other relations with its user, it is no wonder, then, that people have changed their behavior in public. Hans Geser describes the consequences of the never-alone effect combined with quasi-other relations when he observes the scene in Paris' cafes:

> When visiting Paris in summer 2004, after many years of absence, the author was suddenly struck by seeing so few bistro guests reading newspapers, and so many engaged in mobile phone conversation.[12]

Geser's remark mainly concerns the usage of the cell phone for voice communications, and hence is classified here as belonging to the first variation. Such an effect on the world around us remains, and even intensifies, in the next historical variations with the addition of texting, games and e-mails. Today it is evident that newspapers in paper form are in decline, but reading news seems to have moved from the print medium of paper to the digital medium of cell phones (and tablets).

The cafés have changed and so have the streets. About a decade after the launch of the StarTAC, the ability to talk while being away from home or office has made the public phone redundant. The BBC asks almost rhetorically "who uses phone boxes?"[13] Public phones are either severely damaged and not repaired or removed altogether. In 2007, the Washington Post devoted an article to the last public phone booth observed in the Washington, DC area.[14] One of the blogs citing the article comments: "Apparently none of us noticed as these iconic structures slowly slipped away."[15] Now a notice appears at the top of that post: "The phone booth has since been removed." Another blog, "The Payphone Project," reported in 2011 with a sense of surprise: "I made a phone call from this payphone. The call was successful. If you do not use payphones so much these days, then you may have no idea how unusual this is."[16] In some cities, like Paris and London, there were efforts to revive the public phones by turning them into Internet terminals.[17] Time will tell if the effort to "update" the iconic red phone box of London and other public phones will succeed.

Another public asset influenced by cell phones of the first variation is the transport networks. Lyons and Urry suggest travel time is no longer considered wasted time. Hence, they recommend that the public transport system be adjusted to support the new needs of the travelers, whether they travel by car, train or bus. Such adjustments are slowly being made. For example, some airports now offer charging points for cell phone and laptop batteries; and in some airplanes cell phone calls are allowed. On the opposite side, one may find "quiet cars" in trains where phone calls are prohibited. These new codes of behavior create a new ecosystem of traveling and transport that takes into account the ability to communicate while being on the move.

All in all, cell phones of the first historical variation participate in major transformations in the sense of self and the world while contributing to further technological developments. I described here the "never-alone" effect—the diminishing line between home and work, changes in practices performed in public, and the public space in which people move and operate. From a technological perspective, the "seeds" for future developments were identified and will be developed in the next chapters.

StarTAC was chosen to represent the first historical variation because it was well rooted in the mobile voice paradigm: its clamshell configuration brought the microphone closer to the mouth; its size and weight allowed it to be carried everywhere; the holster enabled a quick pickup of a call; the extra battery helped to make more calls while being away; the vibration ensured that most incoming calls were noticed; and the ability to store the phone numbers of interlocutors facilitated calling back. These features and applications improved our ability to talk while away from our home or office. They exemplify how the first historical variation is about mobile speech. The screen's and applications' limited textual capabilities were to develop further in the second historical variation.

NOTES

1. For an analysis of the early history of the cell phone, see Gerard Goggin, *Cell Phone Culture: Mobile Technology in Everyday Life* (London and New York: Routledge, 2006); and Jon Agar, *Constant Touch: A Global History of the Mobile Phone* (Cambridge: Icon Books UK, 2003).

2. Bertolucci, Jeff, 2009, "Star Trek Tech We Use Today (Almost)," *PC World*, May 5, 2009, http://www.pcworld.com/article/164195/star_trek_tech_we_use_today_almost.html (accessed 8 January 2011); Tynan, Dan, "Hello, Gorgeous! 15 of the Greatest Tech Designs Ever," *PC World*, 14 July 2009, http://www.pcworld.com/article/179428/motorola_startac_cell_phone_1996.html (accessed 8 January 2011).

3. Tynan, Dan, "The 50 Greatest Gadgets of the Past 50 Years," *PC World*, December 24, 2005, http://www.pcworld.com/article/123950-2/the_50_greatest_gadgets_of_the_past_50_years.html (accessed 8 January 2011).

4. Mobile Gazette, "Retro: Motorola StarTAC," October 20, 2008, http://www.mobilegazette.com/motorola-startac.htm (accessed 8 January 2011).

5. Ibid.

6. The first short message was sent in 1993, three years before the announcement of the StarTAC, and at the beginning it was considered an "insignificant, additional capability"; see Gerard Goggin, "Mobile Text," *Media Culture* 7, no. 1 (Jan 2004) http://journal.media-culture.org.au/0401/03-goggin.php. This is probably why Motorola presented messages in the user's manual as "pager-like" and "alphanumeric."

7. IEEC (Integrated Electronics Engineering Center), 1996, "Tear Down Digest: Motorola StarTAC Cellular Phone," http://www.ieec.binghamton.edu/ieec/teardown/motorola_star_tac_cellular_phone.htm (accessed 8 January 2011).

8. For cross-cultural research on the attitudes to work and leisure, see Linnda R. Caporael and Bo Xie, "Breaking Time and Place: Mobile Technologies and Reconstituted Identities," in *Machines That Become Us: The Social Context of Personal Communication Technology*, ed. James E. Katz (New Brunswick (US) and London (UK): Transaction, 2006), 219–231. For a sociological analysis of this blurring of boundaries, see Leopoldina Fortunati, "The Mobile Phone as Technological Artifact," in *Thumb Culture: The Meaning of Mobile Phones for Society*, ed. Peter Glotz, Stefan Bertschi and Chris Locke (Bielfeld: Transcript Verlag, 2005), 149–160.

9. Cf. James E. Katz, "Mobile Communication and the Transformation of Daily Life: The Next Phase of Research on Mobiles," in *Thumb Culture: The Meaning of Mobile Phones for Society* by Peter Glotz, Stefan Bertschi and Chris Locke (Bielfeld: Transcript Verlag, 2005), 171–182: 173; Glenn Lyons and John Urry, "Travel Time Use in the Information Age," *Transportation Research Part A* 39 (2005): 257–276; Sherry Turkle, *Alone Together: Why We Expect More from Technology and Less from Each Other* (Basic Books, 2012); Galit Wellner, "Multi-Attention and the Horcrux Logic: Justifications for Talking on the Cell Phone While Driving," *Techné: Research in Philosophy and Technology* 18, no. 1–2 (Winter-Spring 2014): 48–73.

10. Rich Ling and Leslie Haddon, "Mobile Telephony, Mobility, and the Coordination of Everday Life," in *Machines That Become Us: The Social Context of Personal*

Communication Technology, ed. James E. Katz (New Brunswick and London: Transaction, 2006), 245–265.

11. Fortunati "Technological Artifact," 154–155.

12. Hans Geser, "Is the Cell Phone Undermining the Social Order? Understanding Mobile Technology from Sociological Perspective," in *Thumb Culture: The Meaning of Mobile Phones for Society*, ed. Peter Glotz, Stefan Bertschi and Chris Locke (Bielfeld: Transcript Verlag, 2005), 23–35: 24.

13. Geoghegan, Tom, 2008, "Who Uses Phone Boxes?" *BBC News Magazine*, http://news.bbc.co.uk/2/mobile/uk_news/magazine/7228184.stm (accessed 28 April 2011).

14. Deane, Daniela, 2007, "Step Inside, and Back in Time, and Dial Away," *Washington Post*, November 22, 2007, http://www.washingtonpost.com/wp-dyn/content/article/2007/11/21/AR2007112102363.html (accessed 28 April 2011).

15. Twelve Mile Circle, 2007, "Washington DC Area's Last Phone Booth," http://www.howderfamily.com/blog/washington-dc-areas-last-phone-booth/ (accessed 28 April 2011).

16. The Payphone Project, 2011, "A Rare Fully-Functioning Payphone," April 25, 2011, http://www.payphone-project.com/ (accessed 28 April 2011).

17. Paris Invisible, 2010, "The Rebirth of the Phone Boot," http://parisisinvisible.blogspot.com/2010/05/rebirth-of-phone-booth.html. May 2010 (accessed 28 April 2011); Icons.org, "History of the Red Phone Box," http://www.icons.org.uk/theicons/collection/the-phone-box/biography/the-red-phone-box-biography-finished (accessed 28 April 2011).

Chapter 2

Second Historical Variation

Texting-at-Hand

NOKIA'S 5110 (MARCH 1998)

In March 1998, Nokia announced the 5110 handset, whose claim to fame was the ability of users to change the plastic cover of the cell phone "without any special tools." This, the company claimed, "highlights Nokia's commitment to the values of individuality, innovation . . . and human touch." It was the first colorful cell phone.

The 5110 also had applications, although they were not termed as such. These included a calculator and three games—Memory, Snake,[1] and Logic. The games required a larger screen for their operation, and so the 5110's screen was conceived to be relatively large: 84 by 84 pixels that accommodated five lines, four of which were intended for text display. It was a significant improvement compared to a single-line ten-digit screen of the StarTAC.

The 5110 was a GSM cell phone, meaning that the communications between the handset and the antenna were digital. The GSM standard allows the transmission of some text in parallel with voice, also known as Short Messaging Service (SMS). The SMS boom occurred in 1998,[2] the year of the 5110's launch. Since the 5110 was designed and developed months and years before, it can be regarded as occupying the border area between voice and text, with "one leg" still in the voice paradigm and the other already in texting. Yet, the press release announcing the 5110 barely mentions its non-voice capabilities. These new features could be easily overlooked due to the voice paradigm that reigned at the date of the announcement. At that time, messaging was emerging, and it was difficult to forecast its popularity and general public acceptance.

THE USER'S PERSPECTIVE

The Fashion Phone

Nokia made some effort to position the 5110 as "fashionable" and "friendly." For example, when the device was turned on, a welcome notice appeared: "Have a Good Day." It was a handset that called for empathy. Indeed, the users embraced the 5110 and were fascinated by the colored covers and the games, as reflected in the memories of a commentator in one of the forums:

> I would like to hereby decree that the Nokia 5110 was the first ever cool mobile phone. With the interchangable covers and the inclusion of the Snake game it was too cool for school. I remember pestering my parents relentlessly to get them to buy one for me - and when they did I was a legend at my school. However - it was only then that I realised that nobody else I knew had a phone and I couldn't call or text them . . . whoops.[3]

This excitement was not limited to teenagers. The *New York Times* and the *BBC* adopted Nokia's terminology and called it a fashion phone.[4, 5]

Being fashionable usually requires some work, and the 5110 is no exception. Certain marketing and technical materials in the French language reveal the difficulties in changing the colored cover; it was not as simple as the company's announcement made it seem. In practice, the change could hardly be performed "without any special tools," and required a coin, according to the French collaterals, "and some brute force" as testified by one of the users.

In spite of such difficulties, the colored covers were significant enough to lead computer industry analysis firm Gartner to publish a report a year after the announcement of the 5110 in which it determined that the external design features were a key to users' adoption of a new handset:

> A handset's physical characteristics include its appearance and ease of use. While a handset's look makes a statement about the user, its ease of use increases one's comfort level with a product. Vendors, therefore, need to address both. Nokia has been a leader in enhancing the physical appearance of the phone, both through innovative snap-on/snap-off covers and embedded antennae. Other vendors will have to follow Nokia's lead because it is one of the factors that will help them safeguard their market position.[6]

Gartner recommended that cell phone manufacturers invest in enhancing the external appearance of the handset in order to increase its attractiveness for users. Interestingly, the analyst's report did not refer to the voice quality and other auditory traits. This lacuna indicates that the cell phone was no longer perceived as an auditory-only device. It was a technology whose embodiment relations with the user were a mix of auditory and visual capabilities. Gartner's recommendation for external visual improvement encompassed an understanding that a handset was being looked at, for example, when playing games or texting. Bystanders also looked at the handset. Nokia's focus on the changeable covers positioned the 5110 as a fashion phone. As a fashionable product, the color and shape of the handset became major factors in the buying process.[7]

The Cell Phone's Menu

At the time of the launch, the primacy of voice was paramount. This is reflected in the User Guide that reviews the cell phone's functions in their historical order. The first is "Basic Functions" which explains mostly voice-related activities. Next, under the headline "Other Essential Functions," the User Guide describes "Selecting a Ringing Tone." Then it goes through "Voice Messages," and only after that describes "Text Messages" before it ends with "Keypad Lock."[8] Out of four functions, two are directly related to voice. The following chapter, "Using the Menu," repeats the previous two chapters with some extra details (menu options #1–5), thereby fixating on the primacy of voice. Only afterward does the User Guide refer to the games and the calculator (menu options #6 and 7), even though they were mentioned in the Nokia 5110 press release as key innovations. The last menu option is the clock (menu option #8). The order in which the features of the 5110 were listed, both in the handset's internal menu and in the documentation, demonstrates the importance that Nokia (and many users) placed on voice communications over other functions, leaving the company faithful to its slogan "Connecting People."

The primacy of voice was so conclusive that texting was considered a relatively limited service. The User Guide specifically warns: "Make sure before sending a text message that the receiving phone is provided with SMS capability." As a relatively young and immature service, SMS required some checks before being used. Yet, this single warning is an improvement compared to the three conditions mentioned for using StarTAC's messaging service.

As voice dominated, the 5110's memory capacities were voice oriented, albeit enhanced compared to the first historical variation of the cell phone: the maximum number of records that could be entered in the 5110's phone book was 250 (compared to the 20 that could be recorded in the StarTAC two years before); each record contained a number[9] and a name which could be up to 16 characters;[10] the 5110 remembered the last 8 dialed numbers, last 5 missed calls and last 5 incoming calls.[11] The dominant position of voice pushed non-voice applications to the second half of the handset's menu.

Games were in the sixth entry of the main menu of the 5110, and the description of the three of them altogether (Snake, Memory and Logic) occupied only six lines, compared to a full page of twenty-four lines explaining how to make a phone call, which should have been a trivial action in light of its maturity and frequency of use. There is no explanation in the User's Guide for the relevant keys for the operation of each game and its rules. It is difficult to determine from the User's Guide whether game results could be stored, or whether a game session could be restored and continued. What can this condensation mean? The hypothesis that games were not important should be rejected because they were mentioned in the announcing press release and were placed in the main menu before the calculator and the clock. Is it because the games were easy to operate? It turns out that well-known actions were listed, such as making a call. Still, games were not explained. The minimalistic description of the games probably reflects the marginal status of games compared to voice, but there is no evidential support for this supposition. In the next historical variations of the cell phone, games played an important role and thereby contributed to the growth of the screen. Yet, even in this variation, games like Snake required a screen larger than a single line. Eventually, larger screens encouraged longer texts, and therefore made mobile writing easier.

Next in the handset menu was the calculator. This application used the keypad's numbers in their mathematical meaning, and allowed four basic arithmetical operations. The calculator also served as a currency converter. However, it is not clear from the manual whether the cell phone could remember the last exchange rate entered, or multiple rates and currencies. Unlike the detailed explanation for voice operations, and like the games, the description is minimal. Furthermore, the currency converter was only mentioned deep inside the manual, and was not covered in the highlights provided in the first pages of the manual or in the announcing press release.

Neither was the clock mentioned in the press release. Its entry in the 5110's menu is after the games and the calculator, and like those two features, received only a few lines of description. From the short description it seems that the clock had only one time zone, which needed to be manually reset when replacing the phone battery or when traveling to another time zone. Eventually, this feature became so important that it led people to abandon their watches. The cell phone took over the wrist and pushed away the watch, as described by *The Atlantic* in its November 2010 issue in an article titled "Who wears a wristwatch anymore?"[12] How did such a remote feature take over? The User's Guide does not even refer to the option of using the cell phone as a watch replacement. My guess is that it happened when users set the screen saver to display the time. It was very convenient, especially when time was automatically updated by the cellular network.

The cell phone was not only a watch replacement but also an alarm clock substitute. In the User's Guide most of the instructions referred to the alarm clock's functionality. It would take some more versions for multiple "alarms" to be added and then to connect them to calendar entries as reminders. But relying on the cell phone as an alarm clock turned out to be risky, particularly when the time was automatically updated by the cellular network; a wrong update of daylight savings time produced complaints from users who did not wake up on time or missed meetings because they relied on their cell phones as alarm clocks and watches.[13] These complaints hint at how the cell phone has replaced the alarm clock and highlight the important role it has taken in the everydayness of so many people.

The User's Guide basically explains that as an alarm clock, one single alarm could be set. It can be classified as a "single cell memory," because the memory contains only one datum: the clock needs to remember the local hour, and the alarm needs to remember the wake-up time. If the currency convertor was indeed built for one single rate, it might have fit into the single-cell memory paradigm.

The Birth of the Cellular Writing Machine

Games, Calculator and Clock did not require alphabetic inputs. In the 5110, alphabetic inputs remained confined to SMS and the voice-related address book (where names could be added). Other handsets included additional applications that required text, such as a calendar and a memo. All of these text strings were relatively short, and with the exception of SMS, were intended to store data as entered by the user. Therefore, the second historical variation of the cell phone is characterized by short alphabetic interactions between the user and the handset such as texting (which ends up as interaction with other users), entering meeting and events to the built-in calendar application, maintaining a contact list and taking short notes (to-do's). The emergence of writing on the cell phone was based on (and enabled by) the dual use of the keypad as digits and as letters. Writing with a cell phone entailed the handset becoming a keyboard-like device, where each numerical key of the keypad represented 3–4 letters. Two sub-technologies were aimed at simplifying this form of writing. One was T9, which was designed to reduce the clicks by predicting the intended word.[14] According to Carmen Freher, the technology did not catch on because it required that skilled users stop typing and select the right word from several possibilities. The other sub-technology was a miniaturized keyboard (e.g., Blackberry) in addition to the keypad, which made the handset more similar to a writing machine or a computer.

THE IMPACTS OF THE SECOND HISTORICAL VARIATION

Text Communications

Most research on cellular writing explores the communicative aspects of texting.[15] An interesting question asked in this context is why and when people prefer text to voice.[16] One (convincing) explanation refers to the asynchronous engagement that characterizes SMS communication.[17] Specifically, researchers have discovered that SMS is used when voice communication is perceived as irrelevant, unacceptable or impossible, like interaction with someone who is in a class, a meeting or on an airplane.[18] Other studies have shown that SMS enables communication where it would not have happened as voice. For example, a teenager reported that he asked a friend in a text message how his exam had gone, admitting he would not have called that friend to ask that question, but as a text message this interaction was considered "proper social behavior."[19] Another example is micro- and hyper-coordination for last-minute updates and decisions as per a meeting's exact location and final timing.[20] For example, texting allows one to inform other meeting participants that he or she are about to be late and to adjust the starting time of the meeting in real time. By indicating through a text message that one is about to be late, one is perceived to be responsible rather than simply late. The phenomenon of micro coordination led researchers to conclude that "mobile phones serve to soften one's sense of time . . . The idea of being late may disappear altogether" so that being on time is no longer a merit to pursue.[21] It is a significant shift from the central role of measured time in the modern era.[22] Time becomes fluid, similar to earlier times in history, before people obeyed clocks and watches.

Texting turned out to be many people's preferred mode of communication, even when voice was technically and socially possible. This preference is supported by Nielsen statistics from August 2010, which reflect how young people (teenagers and young adults up to the age of 24) prefer texting to voice. These two adjacent groups extensively use SMS (although the young adults of 18–24 also use voice communications), making SMS "the alpha and omega of mobile communication" for these users.[23] Such usage patterns may imply that this historical variation can also be considered a cultural variation, provided that youth is regarded an independent culture. As Leslie Haddon notes, "it is striking that many of the first wave of national studies of the use of mobile phones by children, teenagers in particular, reported somewhat similar experiences."[24] The intensity of texting by young people crosses geographical boundaries and demonstrates a similar pattern worldwide.

Texting, Embodiment and Hermeneutic Relations

As the handset's external appearance has improved and as texting proliferates, the cell phone is no longer a ready-to-hand that is out of sight, in the pocket or near the ear. When typing, the cell phone "occupies" the hands; it becomes literally *present*-at-hand, and dictates a different embodiment than that required for voice communication.[25] Sadie Plant vividly describes two common embodiments of cell phone users of the second historical variation: the less-skilled users hold the handset in one hand and type with the other "with some force"; the skilled users use minimal gestures, "barely need to look at the keys they use as they make their rapid entries"—in short, they are "nimble and dexterous, even ambidextrous."[26] We can determine from a distance whether a user is skilled in texting by noticing the way he or she holds the cell phone and carries the body towards the device. Also, there is a different gaze for new users: they need to look at the keypad and screen while typing to make sure the right letter is selected, while skilled users look only at the screen to proofread text as it is being written, much like computer screen-keyboard typing practices. The central role of the thumb in skilled users' experience led Plant to announce "the thumb generation," and Goggin "the fifth finger."[27] The thumb generation has a different bodily position which sometimes leads to bodily transformations: the hands are physically changing as demonstrated by the BlackBerry thumb syndrome. It is a form of repetitive strain injury caused by the frequent use of the thumb(s) to press buttons on the miniaturized keyboard of smart phones.

The thumb generation not only has different embodiment relations with the cell phone—it also has different hermeneutics with it. The hermeneutic changes because the text is different. It is not only comprised of words, but also contains new acronyms and new symbols that are forming new signifiers. Due to the widespread use of these new forms, the *Oxford English Dictionary* has recently added LOL (Laughing Out Loud) and OMG (Oh My God) as "official" words, along with the symbol of a heart (to denote love).[28] While acronyms can be easily equated with the function of new words in normal evolution of languages, the symbols—known as "emoticons"—call for another paradigm. I argue that the cell phone's hermeneutic goes beyond alphanumeric discourse with the adoption of emoticons. It is an adoption because emoticons were "imported" to cell phones from Internet forums where their usage is dated back to the early 1980s,[29] well before SMS was developed. They emerged at a relatively early stage of mobile writing because texting technology limits text size to 140 characters, a format whose echoes remain long after the technical barrier was removed in websites like Twitter. Unlike the long texts enabled by digital writing's cut and paste functionality, in the SMS/Twitter world the message is extremely short; it carries no more than a sentence or two, with words rewritten in a much shorter way that encompasses numbers (like "gr8" for "great") or several words "squeezed" into acronyms (like LOL). Emoticons produce an even-greater shrinking effect because they can represent an array of feelings in only three characters. A nine-character sentence like "I'm happy" can be replaced by three characters of colon-hyphen-parenthesis, forming a horizontal smiley—":-)."[30] A similar sort of shrinking occurs when the cell phone's software automatically replaces the three signs with a single-character smiley image—☺. This character has come to carry additional meanings beyond "I'm happy," such as "ha ha ha," "glad to hear the good news," or "thank you for your message." This spectrum of meanings is somewhat similar to the ancient Egyptian pictographic writing that used hieroglyphs.[31] Each hieroglyph could have a complex meaning, beyond a single word. But unlike the Egyptian writing that was exercised by a relatively small group, the cellular emoticons have become popular around the world and across various languages. When I communicate with German colleagues via SMS, a smiley has the same meaning for all of us. Thus, emoticons cross linguistic borders as they are being used by the masses while returning to ancient writing practices of pictorial emotions.

Other aspects of cellular writing also resonate with early forms of writing, such as written lists. The first cellular writing was for contact lists, and then quickly expanded to meetings and notes. I argue that this is not the first time in human history that a writing system kick-starts with lists. According to Jack Goody's "What's in a List," the Sumerian writing system in Mesopotamia (3000 BCE) was developed and used by the bureaucracies of palaces and temples for managing administrative lists. Then it evolved to include additional registers such as botanical and zoological lexicons, as well as astronomical records. Cell phone writing returned to this beginning point and started with lists of phone numbers and their associated names, later followed by more complex lists of schedules and free text notes.

I discuss the *re-turns* to ancient practices because the world and the related technologies have changed so dramatically that a simple straightforward *return* is impossible. One can get a sense of this set of transformations through Ihde's account of the historical variations of writing.[32] First came the "hard" writing techniques of cuneiform, involving a hard instrument like a stick and a hard surface like a tablet. They were followed by "soft" technologies that used brush, quill or pen with paper, parchment or papyrus. The soft tools were held differently than the hard ones and produced different letters. In the nineteenth century the typewriter appeared and implied an altogether different embodiment as two hands were involved in the writing process. Next came computerized writing technologies, which went beyond the hard-soft paradigms. In this case, the text was dynamic in that it could easily be changed on the screen. The cell phone posed an even newer writing paradigm where the thumb interfaces with the technology itself. The different embodiments of each variation of writing dictate different forms for the content. Friedrich Nietzsche complained the typewriter transformed his style of writing. Similarly, Martin Heidegger preferred writing with a fountain pen because it allowed him time to think between the sentences.[33] For mobile writing, complex ideas should be reworked into short messages that can be displayed on the relatively small screen. These short texts should be simple enough to be consumed anywhere anytime. This process is already underway; Twitter university was launched as recently as 2011, and aims at conducting microcourses that consist of 25 tweets (each of 140 characters).[34] This is a different format of knowledge, which possibly answers the floods of new knowledge we confront. Mobile writing, therefore, is about different signs, different embodiments, different formats and probably different content.

The Non-Communicative Writing Machine

Most scholars of mobile writing have referred to emoticons and acronyms as facilitators and accelerators of communication. Other scholars depart from communication themes and instead consider these elements in terms of writing. For example, Maurizio Ferraris constructs the cell phone as "a writing instrument" because it is used for writing short notes, memos, to-do's, calendar entries, or names in a phone book.[35] These short texts have nothing to do with communications with other people. They are all for the benefit of the user. The ability to write a text with the cell phone changed its role from merely "Connecting People," as Nokia's slogan declared, to a writing tool in the sense of Ihde's variations of writing that are evolving from hard to soft to digital media. For Ferraris, writing-as-recording prevails over writing-as-communications, and the keypad is a key feature of a cell phone that records. In writing-as-recording, the cell phone becomes a "memory prosthesis," a human memory that is exteriorized in a technological device. Ferraris' view may explain the rise of cell phones with full keyboards (e.g., RIM's Blackberry or Nokia's Communicator), as well as the increase in memory capacities for the cell phone's various applications. These enable faster, easier and larger recordings of things-to-be-remembered.

In sum, the second historical variation of cell phones is about the transition from voice communications to text. Text is not only for communicating with other people, but also for taking notes, thereby turning the device into a writing tool. The Thumb Generation has a new writing paradigm, where "fingers do the talking" as well as the memorization.[36] In order to write with a cell phone, the user has to look at the cell phone (during the writing process or when proofreading shortly after), which creates new embodiment relations. This new bodily position shifts the attention of users, manufacturers, and designers to the handset's external design. Therefore, this second historical variation is represented by a colorful handset albeit with a black-and-white screen. The alphabetic interactions supported by these cell phones are somewhat limited to texting and address book's names, although it contains simple pictures of the emoticons. The next historical variation is represented by a black handset, with a color-rich screen. It has extensive texting capabilities that allow the composition of relatively long e-mails that go beyond the limits of 140-character short texts. Yet, its focus is not text alone—it is more about multimedia.

NOTES

1. The first Snake game for mobile phones was introduced one year prior, in 1997, for Nokia's 6110 (http://www.nokia.com/about-nokia/company/story-of-nokia/mobile-revolution, accessed 21 September 2010).

2. Gerard Goggin, *Cell Phone Culture: Mobile Technology in Everyday Life* (London and New York: Routledge, 2006), 73.

3. Gdgt.com, 2009, http://gdgt.com/discuss/first-ever-cool-mobile-phone--1ty/ (accessed 15 December 2010).

4. Hafner, Katie, "From Phone to Fashion, Nokia Style; A Designer Who Sets A Worldwide Standard for Technophiles," *The New York Times*, December 9, 1999, http://www.nytimes.com/1999/12/09/technology/phone-fashion-nokia-style-designer-who-sets-worldwide-standard-for-technophiles.html (accessed 11 March 2011).

5. BBC, "See The Evolution of the Mobile Phone from Its Earliest Days," 7 January 2010, http://www.bbc.co.uk/newsbeat/10004114 (accessed 15 December 2010).

6. Jaffery, Naqi A. and Hoffman, Matthew, "Wireless Handsets: The Digital Onslaught," *Gartner*, July 5, 1999, ID Number: MOBD-NA-MT-9901.

7. For a gender-oriented explanation for the emergence of colored and fashionable cell phones, see Leopoldina Fortunati, "Gender and the Mobile Phone," in *Mobile Technologies: From Telecommunications to Media*, ed. G. Goggin and L. Hjorth (New York and London: Routledge, 2009), 23–34: 26. Yet, Fortunati claims that gender differences should be analyzed with care and that one is likely to find a "more complex picture" that reveals a not-so-clear-cut picture.

8. Keypad lock was required for a candybar configuration of a handset to prevent the keys of the keyboard from being accidentally pressed when the handset is stored in a pocket or a handbag.

9. There is no reference in the User Guide to a limit in the number of digits.

10. These phone numbers and names could be used also for texting. Yet, it was not the main purpose of the phone book, as can be evidenced by its name (*phone* book) and the order of the presentation of the phone book in the user manual as part of chapter 3, "basic functions," where the voice functionality is detailed.

11. Interestingly, some of these memory capabilities were reduced, compared, for example, to the 8110 that remembered last ten numbers in each category. A possible explanation may be that a more "durable" memory in the framework of an address book reduces the need to remember last dialed number or incoming calls, as there is a probability the number is already stored with a name associated with it. Put differently, there is as a conversion of short-term memory (last calls) with long-term (address book). This can be an important step for the cell phone to become a significant player in our everydayness that stores necessary information for us.

12. See "Who Wears a Wristwatch Anymore?" in the November 2010 issue. (http://www.theatlantic.com/magazine/archive/2010/11/a-new-wrinkle-in-time/8249 (accessed 27 October 2012).

13. See Graham Harman's blog post of 29 April 2011on iPad's failure to update time in Egypt in 2011. http://doctorzamalek2.wordpress.com/2011/04/29/an-hour-earlier-than-i-thought/ (accessed 24 June 2013); Horn Leslie, 2011, "iPhone Clock Fails to Adjust to Daylight Savings Time," *PC Magazine,* http://www.pcmag.com/article2/0,2817,2381901,00.asp (accessed 24 June 2013) on iPhone's wrong update of daylight savings in the USA; or Blackberry's wrong change in Israel in October 2010 described in Ruti Levi, 2010, "Blackberry Updated by Global Daylight Saving Clock, Many Woke Up Late," *Haaretz,* http://192.118.73.5/hasite/spages/1191861.html (accessed 24 June 2013).

14. For the analysis of the application of a known interface, such as a keyboard, to a new context, such as a keypad, see Robert Rosenberger's notion of "abstract relational strategies" in his article "Embodied Technology and the Dangers of Using the Phone While Driving," *Phenomenology and the Cognitive Sciences* 11, no. 1 (March 2012): 79–94.

15. For example, Heike Weber, "Consumers as Innovative Actors? The Role of Users in the Shaping of German GSM telephony," *Le Mouvement Social* 3 (2009): 117–130. Rich Ling and Birgitte Yttri, "Hyper-Coordination Via Mobile Phone in Norway," in *Perpetual Contact: Mobile Communication, Private Talk, Public Performance*, ed. James E. Katz and Mark Aakhus (Cambridge: Cambridge University Press, 2002), 139–169; Rich Ling and Scott W. Campbell, *The Reconstruction of Space and Time: Mobile Communication Practices* (New Brunswick: Transaction Publishers, 2009); Paul Levinson, *Cell phone: The Story of the World's Most Mobile Medium and How It Has Transformed Everything!* (New York and Basingstoke: Palgrave MacMillan, 2004).

16. For example, Rich Ling, *The Mobile Connection: The Cell Phone's Impact on Society* (Morgan Kaufman, 2004), 147–148; Goggin, *Cell Phone Culture*, 74–88.

17. Cf. Jason Farman, "Mobile Media Performances as Asynchronous Embodiment," *The International Hournal of Screendance* 2 (Spring 2012B), 48–51.

18. An uncommon example for using texting when voice is impossible was provided my late aunt Zehava, who suffered for two years from Amyotrophic Lateral Sclerosis (ALS), also known as the Lou Gehrig' disease. In her case, the first functions she lost were that of her right arm and her speech. She talked to us via her cell phone, which she used as a writing pad. It was communication with a cell phone that did not require the cellular network. Yet, it made communication feasible in circumstances in which it was hardly possible otherwise.

19. Louise Barkhuus, "Why Everyone Loves to Text Message: Social Management with SMS," *GROUP'05* (Sanibel Island, FL, USA, 2005), 324–325.

20. Ling and Yttri, "Hyper-Coordination."

21. James E. Katz, "Mobile Communication and the Transformation of Daily Life: The Next Phase of Research on Mobiles," in *Thumb Culture: The Meaning of Mobile Phones for Society*, ed. Peter Glotz, Stefan Bertschi and Chris Locke (Bielfeld: Transcript Verlag, 2005), 171–182: 172. See also John Urry, *Mobility* (Cambridge: Polity, 2007), 173.

22. Lewis Mumford, *Technics and Civilization* (Harcourt, 1934); Linnda R. Caporael and Bo Xie, "Breaking Time and Place: Mobile Technologies and Reconstituted Identities," in *Machines That Become Us: The Social Context of Personal Communication Technology*, ed. James E. Katz (New Brunswick and London: Transaction, 2006), 219–231: 220.

23. Ling, *The Mobile Connection*, 145; see also Goggin, *Cell Phone Culture*, 74; Leopoldina Fortunati, "The Mobile Phone as Technological Artifact," in *Thumb Culture: The Meaning of Mobile Phones for Society*, ed. Peter Glotz, Stefan Bertschi and Chris Locke (Bielfeld: Transcript Verlag, 2005), 149–160: 156.

24. Leslie Haddon, "Looking for Diversity: Children and Mobile Phones," *An International Conference on Social and Cultural Aspects of Mobile Phones, Media, and Wireless Technologies* (Australia: The University of Sydney, 2007).

25. Cf. Howard Rheingold, *Smart Mobs: The Next Social Revolution* (Cambridge: Basic Books, 2002).

26. Sadie Plant, *On the Mobile: The Effects of Mobile Telephones on Social and Individual Life* (Motorola, 2003), http://www.motorola.com/mot/documents/0,1028,333,00.pdf, 53.

27. Gerard Goggin, "Mobile Text," *Media Culture* 7, no. 1 (January 2004): http://journal.media-culture.org.au/0401/03-goggin.php.

28. http://www.oed.com/public/latest/latest-update (accessed 4 April 2011); see also coverage in CNN - Stewart, Anne, "OMG! Oxford English Dictionary adds new words," *CNN*, 25 March 2011, http://articles.cnn.com/2011–03-25/living/oxford.new.words_1_new-words-oxford-english-dictionary-usage?_s=PM:LIVING (accessed 4 April 2011).

29. Alecia Wolf, "Emotional Expression Online: Gender Differences in Emoticon Use," *Cyber Psychology and Behavior* 3, no. 5 (2000): 827–833.

30. For variations of the smiley see Carmen Freher, *Email - SMS - MMS: The Linguistic Creativity of Asynchrnonous Discourse in the New Media Age* (Bern: Peter Lang AG, International Academic Publishers, 2008).

31. Cf. Don Ihde, *Postphenomenology: Essays in Postmodern Context* (Evanstone: Northwestern University Press, 1995), 84.

32. Don Ihde, *Heidegger's Technologies: Postphenomenological Perspectives* (New York: Fordham University Press, 2010), 129–139.

33. Friedrich Kittler, *Gramophone, Film, Typewriter*, trans. Geoffrey Winthrop-Young and Michael Wutz (Stanford: Stanford University Press, 1999).

34. http://svtwuni.wordpress.com/2011/12/07/stu06/ (accessed 1 March 2012).

35. Maurizio Ferraris, "Where are you? Mobile ontology," *Seeing, Understanding, Learning in the Mobile Age* (Budapest, 2005) http://www.fil.hu/mobil/2005/Ferraris.pdf (accessed 22 October 2010).

36. Levinson, *The Story*, 116.

Chapter 3

Third Historical Variation

The Kingdom of Multimedia Applications

APPLE'S iPHONE (JANUARY 2007)

The third historical variation is characterized by an abundance of applications, like weather forecasts, stock market data, maps, size converters, dictionaries, music players, games, mobile TV, mobile Internet access, or put literally—"multimedia." The phone is no longer named "cell phone," "cellular phone" or "mobile phone," but rather "smart phone." Within this sphere, the iPhone is the poster child of smart phones, named by Time Magazine as one of the best inventions of 2007.[1]

The iPhone was Apple's first cell phone,[2] composed of a large screen and one single button. When announced in January 2007, both the initial press release and the presentation by the company's CEO Steve Jobs were extremely upbeat. In the press release Jobs praised the iPhone as not only "revolutionary"[3] but also a "magical product" and "literally five years ahead of any other mobile phone." At the unveiling of the iPhone, Jobs stated: "Today Apple is going to re-invent the phone!" And when he demonstrated at that event how the iPhone works, the audience expressed excitement and astonishment by applauding and yelling after every feature demonstrated. Jobs encouraged such responses by uttering "I could play with this for long time" (the audience laughs), "isn't that awesome?," and "pretty cool, ha?" *Wired Magazine* enthroned that event as "the biggest launch since the Apollo program."[4]

This enthusiasm endured for more than six months and until the device was launched—that is, until it became available for sale. *New York Times*' columnist David Pogue described this period in terms of hype:

> Talk about hype. In the last six months, Apple's iPhone has been the subject of 11,000 print articles, and it turns up about 69 million hits on Google. Cultists are camping out in front of Apple stores; bloggers call it the "Jesus phone." All of this before a single consumer has even touched the thing.[5]

As a journalist for the *New York Times*, Pogue was eligible to receive an iPhone before it hit the market and experience it before the masses would get their hands on it. Although his pre-launch article on the iPhone opened with a cautious tone, Pogue quickly turned passionate, and enthusiastically wrapped up his review by taking back the "hype charge:"

> But even in version 1.0, the iPhone is still the most sophisticated, outlook-changing piece of electronics to come along in years. It does so many things so well, and so pleasurably, that you tend to forgive its foibles. In other words, maybe all the iPhone hype isn't hype at all.

Pogue attempts to rationalize the excitement around the iPhone and at the same time justify it. When the iPhone was finally available six months later, it turned out that the public was just as enthusiastic as the columnists, opinion leaders and industry gurus.[6] Even a year later the excitement still had not worn off, and *Wired Magazine*'s contributing editor Fred Vogelstein said, "The hosannas[7] greeting the iPhone were so overwhelming it was easy to ignore its imperfections."[8]

THE USER'S PERSPECTIVE

Putting aside the marketing, branding and consumer expectation management mentioned in the previous section, the iPhone's innovation can be boiled down to two focal points—the touch screen and the plethora of applications.[9]

Apple's press release for the launch of the iPhone described the device as having a "stunning 3.5-inch widescreen display" capable of displaying videos and browsing "any web page the way it was designed to be seen, and then easily zoom in to expand any section by simply tapping on iPhone's multi-touch display with [the] finger." Apple was proud of the large size of the screen (which was further extended in the iPad a few years later). But it was not just a larger screen; it was a touch screen. A touch screen means that it is not only viewed, but also touched, thereby dictating a new interface with the user, or in postphenomenological terms—a different embodiment. The phone is operated not by pushing buttons, but instead by touching the screen, usually with the pointing finger. Steve Jobs was quoted in the press release as saying: "We are all born with the ultimate pointing device—our fingers—and iPhone uses them to create the most revolutionary user interface since the mouse." While the second historical variation of the cell phone was based on the thumb, users of a touch screen cell phone employ the pointing finger (with the addition of the thumb in cases when two fingers are needed, such as zooming in and out). The pointing finger not only presses buttons and keyboards presented on the screen, but also browses over lists, selects, and zooms in and out. That is why the iPhone was designed with only one button– a big difference compared to cell phones that have keypads, or smartphones that have miniaturized keyboards (e.g., Blackberry handsets). Gerard Goggin calls it "a haptic turn."[10]

With a touch screen, the user is presumably physically closer to the cell phone because he or she uses their fingers to interface with the cell phone instead of more traditional pointing techniques like a mouse or a stylus. Yet, the name "touch screen" should not be confused with the sense of touch. A truly haptic technology would enable a user to actively feel and interact in a different way from that offered by a keyboard. From this point of view there is no significant difference between the virtual keyboard that Apple's iPhone features or the miniaturized keyboard featured by Blackberry, Nokia and other cell phones. The fact that a touch screen requires the usage of a tactile organ (i.e., the fingers) does not turn it into a haptic technology. The screen remains smooth and sleek, regardless of the application and the content.

If both keypad and touch screen convey equivalent haptic experiences, can one argue that there is no change in the embodiment? After all, Motorola's StarTAC (the first historical variation) and Nokia's 5110 (the second historical variation) also interfaced with the user's finger. The only difference would be, according to this claim, that the finger touches the keypad instead of the screen. By looking at the details of the interfacing techniques and their related embodiments, the difference becomes evident. Let's take for example a relatively simple task like selecting a phone number from an address book. For cell phones like StarTAC and 5110, this operation requires pressing the up and down keys until the target contact is reached, and then pressing another key for the OK/enter/select action. Unskilled users need to switch their gaze from the screen where the list is displayed to the keys to be pressed. In contrast, this operation on the iPhone requires having the fingers on the surface of the screen, leaving the gaze fixed on the list (because it is presented on the screen). The scrolling and selection are done with the pointing finger that is moving on the list displayed on the screen. That was considered revolutionary. New York Times' Pogue described the demonstration of a similar operation (selecting from a list of music items) in the following words: "The crowd went nuts when Steve Jobs demonstrated how you scroll through your iTunes music list: you flick your finger upward or downward on the screen. The list flashes by, slowly coming to a stop like a roulette wheel."[11]

Basic operations are surprisingly displayed on the iPhone's screen, producing a kind of technological enchantment.[12]

As noted above, the hype around the iPhone had two foci: its screen and its applications. The screen was praised for its size and qualities as a touch screen; but only three applications were mentioned in the press release. Apple described the iPhone as a combination of three products—a mobile phone, an iPod, and an Internet communications device. For Apple, the combination of three technological devices into one was the major novelty of the product.[13] However, when the iPhone was eventually released to the public, it initially came with a set of fourteen applications.[14] The user could arrange them in any order, but could not add any new ones; the App Store had not yet been announced.

Interestingly, of all applications, the first feature set described in the press release was voice communications: how to make a call, how to operate the address book and how to use the speed dial. Apparently it was not easy to abandon the voice heritage of cell phones, even though Apple tried to redefine the concept of cell phone and voice communications practices.

However, the strength of the iPhone is not necessarily in its voice capabilities. In the words of the *Consumer Reports*: "telephony may not be its strong suit."[15] As an owner of an early model of the iPhone I can attest that the voice quality was not as superb as the screen. It was the victory of the visual over the auditory. It frequently made me wonder why the iPhone's name even contains the word "phone."

Sooner or later the application paradigm took over, and voice communication and texting became applications among many others. The iPhone's applications have been characterized by their abundance and breadth of topics. In July 2008, a year after the iPhone's launch, the number of applications dramatically increased when Apple opened the "App Store." The App Store is an online database that functions as a virtual store where iPhone users can download applications for their device for a few dollars or for free. Initially, the store was launched with 500 applications.[16] One user commented on a news item that covered the launch of the App Store: "125 free apps? How am I supposed to fit all of those on my phone?"[17] It seems that even the initial quantity of 500 applications was beyond a single user's ability to experience. It was a major change from other cell phones in which all applications of a given device could be handled. The number of applications for the iPhone has grown exponentially since then: from 500 applications in July 2008, to over 100,000 by November 2009, and 425,000 by July 2011.[18] These numbers do not cover all of the applications available for iPhones. More can be accessed in Cydia, a platform for downloading applications for iPhone that is not controlled by Apple.[19] Because Cydia is based on open source, it is difficult to know how many applications are available there.

THE IMPACTS OF THE THIRD HISTORICAL VARIATION

The Quasi-Other

While the second historical variation was focused on writing, that is—on the *input* of text through a keypad, the third historical variation of the cell phone has moved to *output*. Its major function is the "exteriorization" of information stored in a cell phone or on the Internet. For handsets of the third variation, the center stage is no longer occupied by the keypad or by a miniaturized keyboard. It gives way to the screen. Users look at the screen to read e-mails, select camera shots, review a map, or type on the touch screen that displays a full keyboard. A physical keyboard is optional, not a must.

The new screen brought a new type of entertainment. In the App Store there is one group of applications that is classified in the category of games and entertainment, but does not follow the traditional game layout that requires the user to follow a certain logic in order to reach some targets. Instead, these applications present a fictitious character that serves as a virtual pet. For example, in an application named "Talking Tom Cat,"[20] a drawn cat is presented on the screen. The cat has an arsenal of responses to touch, and repeats what he "hears" in a funny voice. Within this category one may find additional virtual pets like a cow, a giraffe, a dinosaur, a bacterium and other creatures. All these produce alterity relations,[21] turning the cell phone into a friendly (and funny) quasi-other. Such entertainment applications exemplify how technologies can become present-at-hand in a positive scenario, in contrast to the Heideggerian schema in which technologies come to the foreground due to malfunctioning, breakup or absence, or as subjects of a scientific investigation.[22] Talking Tom Cat and his "friends" occupy the foreground in and of themselves and in a positive context of properly working technologies. I am aware of the anthropomorphic direction this analysis might be taking,[23] but I wish to show how such an examination can enhance the understanding of the cell phone experience. My claim is that Talking Tom Cat helps us refer to the handset as a "fellow" or a companion so that the cell phone becomes a quasi-other with which we maintain alterity relations.[24]

The quasi-other not only entertains, it also provides useful information. According to the findings of a study of the cell phone habits of new mothers published by BabyCenter in its "21st Century Mobile Mom Report,"[25] new mothers prefer cell phones with cameras and applications. The report also found that these new mothers mostly use applications that are classified as "for kids" and "health and wellness." The selection of these types of applications means that new mothers do not only get advice from talking to relatives and neighbors, or from reading the legendary Dr. Spock's *Baby and Child Care* and its successors, but also from websites, blogs and social media.

The mediated other (who may be another new mom in a forum) is considered by new mothers as a reliable and trustworthy source of know-how. Additionally, unlike access via a computer that requires the user to sit and devote significant attention, access via a cell phone can be done while moving or being occupied in other activities, such as feeding the baby. As a result, the cell phone becomes so central to these new mothers that "the smartphone screen is more intimate than any other medium . . . the majority of moms sleep with their cell phone next to their bed . . . more than half check their phone first thing in the morning and last thing before they go to sleep . . . it's her constant companion that makes juggling easier . . . "[26] The report asserts that because of the centrality of the cell phone in the new mothers' lives, a new motherhood is produced. It is the result of a mother-cell phone combo in which the mother maintains alterity relations with the cell phone and the cell phone becomes a quasi-other. This quasi-other is a companion that entertains and advises.

Customization, Personalization, Domestication

The "cellular" quasi-other is now different for each user because a cell phone of the third historical variation can be extensively customized according to personal needs and tastes. This is a new level of customization since users can add or remove functionality. Conversely, customization for cell phones of the first historical variation was done mostly by using stickers and hangers to externally differentiate an individual handset from the mass. When the stickers included pictures (of the owner, her friends, or his children) or when the hangers were handmade, customization turned into personalization, meaning that nobody else had one "like this." By the second variation, external customization was enhanced by the ability to flip covers. More importantly, basic internal customization emerged and enabled downloading ringtones and choosing graphics for the opening screen or for a "screen saver." It was also possible to personalize by setting "quick dial" numbers.[27] For the first and second variations, most of the adjustments were "cosmetic" because the basic functionality was set by the manufacturer. One of iPhone's key innovations was to allow users to add applications, remove others and rearrange them altogether. The iPhone's customization has reached such levels of variety that it becomes personalization, for even after a short period of use, two iPhones would not be identical.

These practices of customization and personalization fit into Roger Silverstone's sociological notion of "domestication," originally developed for home technologies such as the television and the personal computer. Domestication generally means "taming the wild," and in the context of technology it refers to the ongoing process of transforming a technological object "acquired from the public domain" and making it "a part of the home."[28]

As a sociological notion, it deals with the relations between the members of the household as they are reflected in their consumerist attitudes toward a technological artifact. This notion was later extended to cell phones by Leslie Haddon. He notes that the domestication of cell phones might be considered paradoxical because cell phones are mostly used outside of the home, while domestication implies a strong connection to the home and the activities conducted within this domain. Haddon solves this paradox by a virtual extension of the home to incorporate the out-of-home spaces where the cell phone is typically operated. As Paul Young sang, "wherever I lay my hat that's my home." Domestication adds to customization the effect of feeling-at-home with technological artifacts, and weaves these artifacts into everydayness.

The outward extension of the home may be necessary from a sociological perspective, but it can be bypassed. For example, a postphenomenological analysis is likely to investigate the relations between users and technologies in and of themselves, while leaving the home as one possible background. Yet, domestication theory can enrich the postphenomenological scheme of I-technology-world. Once we equate users and consumers, domestication adds layers to the scheme—the individual object (i.e., my handset), the product line (i.e., Nokia 5110) and the generic technology (i.e., cell phones).[29] Domestication can help frame the differences between the cell phone in-actual-use on one hand, and cell phone à-la-user-manual or the cell phone as described in the press releases on the other. In other words, the notion of domestication marks the consumerist demarcation point between the generic and the specific object, and places a technological object before any action of buying, customization and personalization. Our relations with our cell phones can start as early as the planning stage before buying the specific handset, thereby extending the range of their mediation to a stage before the actual usage.

The concept of domestication may explain the various changes one makes in her handset, but it is more difficult to contain the nuance of hyper-personalization within this concept, as evidenced in the third historical variation of the cell phone. The intense urge to customize the iPhone with applications can be conceptualized with Ihde's notion of "drastic multistability" that characterizes contemporary digital technologies (DTs). Traditional tools and machines, like a hammer, a car or a book, are multistable because they have different meanings to different people at different times.[30] However, the practices associated with traditional technologies are rarely followed by an ambition to change them or improve them, at least not on a wide scale as exhibited by the iPhone (and its followers). Changes users make to traditional technologies mostly consist of personalization at the level of the single specific artifact, such as an *ex libris* stamp on books. By contrast, DTs call for a major adaptation to the user's specific tastes and needs at the product level, thereby intensifying the multistability of the individual objects. Put differently, while every hammer can serve as a paper weight, each cell phone can be customized to a different usage, such as a communication device, a navigation system, music player, a clock, a weather forecast terminal, and more. The result is drastic multistability. Combining multistability and domestication creates a richer array of changes that take place in various points of the artifact's life cycle, while providing a level of dynamism that is necessary for fast-changing technologies such as the cell phone.

Everyday Carry

The drastic multistability of the cell phone of the third historical variation results in the elimination of many "Everyday carry," that is—the things we hold in our bags, pockets, and purses and which differ from one person to another.[31] This phenomenon was noticed by *TechCrunch* editor John Biggs:

> Who needs a Leica when we carry an iPhone? Who needs a notepad when we carry a Nexus S? Who needs a book when we carry an iPad? These devices don't have the same import, the same heft as their simpler counterparts but does that really matter? They do the job well enough or better.[32]

The writing capabilities of the second historical variation allowed us to replace our notebooks, schedules and watches with our cell phones. Now we can also leave behind children's pictures, a compass, paper tickets for a show, a concert or a flight (replaced by a text message containing a barcode which approves the purchase of "tickets"), or even our wallets (replaced by m-payment) and car keys.[33] It becomes the new "Swiss army knife," even beyond the humorous illustration on the cover of Paul Levinson's book *Cellphone*. Cell phones of the third historical variation transform some of our everyday practices by replacing objects that were around us for hundreds of years with a digital "virtual" functionality. Keys, maps, tickets, compass, flashlight, all these (and more) become redundant if you have the right application installed on your cell phone.

This change in the world of artifacts is accompanied by a change in users' practices. While traditional everyday carry requires a certain anticipatory planning, the ability of cell phones of the third variation to provide constant access to a wealth of data allows us to be more spontaneous and plan our actions less in advance. We no longer need to predetermine which restaurant to go to, which book to read on the way, or which music to listen to. Selection is done on the spot, instead of at home, in advance, before we go. The cell phone becomes the enabler of a new spontaneity.

In sum, the shift from the second historical variation to the third moved the focus on cell phones from *input* to *output*. The keypad and keyboard gave way to the screen, which became the dominant component. In the fourth historical variation, the pendulum moves back to input as the main focus. This time, input devices are multiplied so that the keyboard becomes yet-another input device, along with the cell phone's microphone, camera and GPS.

NOTES

1. Grossman, Lev, "Invention of the Year: The iPhone," *Time,* November 1, 2007, http://www.time.com/time/specials/2007/article/0,28804,1677329_1678542_1677891,00.html (accessed 13 November 2012).

2. The other products were computers and IT-related devices. Apple was involved in other cell phone projects, but in collaboration with other manufacturers, see Gerard Goggin, "Adapting the Mobile Phone: The iPhone and its Consumption," *Continuum: Journal of Media & Cultural Studies* 23, no. 2 (2009): 231–244.

3. An academic analysis offers a more balanced approach and sees this product launch as an evolution rather than a revolution; see Goggin, "Adapting the Mobile Phone," 235.

4. http://www.wired.com/culture/geekipedia/magazine/geekipedia/iphone, accessed 24 June 2013.

5. Pogue, David, "The iPhone Matches Most of Its Hype," *New York Times*, June 27, 2007, http://www.nytimes.com/2007/06/27/technology/circuits/27pogue.html?_r=1&ref=iphone (accessed 5 April 2011).

6. The first weekend's sales exceeded expectations of 400–500,000 units, and reached 700,000 units (Riley, Duncan, "iPhone Weekend One: 700,000 Sold, $200 million+ Profit For Apple," TechCrunch, July 4, 2007, http://techcrunch.com/2007/07/04/iphone-weekend-one-700000-sold-200million-profit-for-apple/, accessed 13 December 2010). Within 74 days, Apple sold one million units of iPhone (http://www.apple.com/pr/library/2007/09/10iphone.html, accessed 13 December 2010).

7. Hosannas is originally a Hebrew word which was adopted by contemporary English language, meaning "A Shout of Fervent and Worshipful Praise," according to American Heritage Dictionary.

8. Vogelstein, Fred, "The Untold Story: How the iPhone Blew Up the Wireless Industry," *Wired*, 1 September 2008, http://www.wired.com/gadgets/wireless/magazine/16–02/ff_iphone?currentPage=4#ixzz180Uq36O9 (accessed 13 December 2010).

9. The screen was already identified in 2007 as a key innovation by *Time* magazine (see Grossman in fn. 1). Applications for cell phones were available before the iPhone's launch. For example, there were applications to buy soft drinks from a machine, pay for a ski pass or groceries (see Adriana de Souza e Silva, "From Cyber to Hybrid: Mobile Technologies as Interfaces of Hybrid Spaces," *Space and Culture* 9,

no. 3 (August 2006): 261–278, 272). The novelty of the iPhone was the easy access through the App Store. No wonder, then, that the App Store concept was duplicated by Google for its Android Google Market, by Nokia for its platforms and by other cell phone vendors.

10. Goggin, "Adapting the Mobile Phone," 242.

11. Pogue, David. 2007B. "The iPhone Rumors Are Right . . . Finally," *New York Times*. January 9, 2007, http://bits.blogs.nytimes.com/2007/01/09/the-iphone-rumors-are-rightfinally/?ref=iphone (accessed 5 April 2011).

12. These operations look simple and effortless, as though they urge the users to obey the (humorous) rule formulated by Wired magazine: "Never Read the Manual First" (Wired. 2009, "How to Behave: New Rules for Highly Evolved Humans," *Wired Magazine* 17.08 http://www.wired.com/culture/lifestyle/magazine/17–08/by_index (accessed 13 December 2010)). Yet, iPhone's User's Guide has 217 pages (see Apple, "iPhone User's Guide." http://manuals.info.apple.com/en/iphone_user_guide.pdf, accessed 24 June 2013), compared to 55 pages of Nokia's 5110, and 126 pages for Nokia's 9000i (Communicator) which is the first smartphone launched in 1996.

13. Interestingly, until the iPhone, it was believed that mobile Internet could not take off fast because "person-to-person functionality is valued more highly than person-to-information functionality" (Jane Vincent, Leslie Haddon, and Lynne Hamill, "The Influence of Mobile Phone Users on the Design of 3G Products and Services," *The Journal of the Communications Network* 4, no. 4 (October–December 2005): 69–73, 70). As the iPhone proliferated, the ability to access the Internet through a cell phone became more popular (see Goggin, "Adapting the Mobile Phone"). With a suitable technology, people change their mode of behavior and leave behind deterministic statements, which were considered correct for a previous technological generation. Jason Farman notes that by 2009 cell phones were used more for data (that is—texting and Internet access) than traditional voice communications, see his article "Mobile Media Performances as Asynchronous Embodiment," *The International Hournal of Screendance* 2 (Spring 2012): 48–51.

14. These included a phone application (with embedded contact list), mail, web browser (Safari), media player (iPod), texting, Calendar, photo & camera, video player (YouTube), Stocks, Google Maps, Weather, Clock, Calculator, and Notes.

15. Gikas, Mike, 2007, "iPhone, the Cell Phone," *Consumer Reports*, 3 July 2007, http://blogs.consumerreports.org/electronics/2007/07/iphone-the-cell.html?CMP=I72RSE0 (accessed 17 January 2011). See also Ian Bogost, *The Geek's Chihuahua: Living with Apple* (Minneapolis: University of Minnesota Press, 2015).

16. Ricker, Thomas. 2008. "Jobs: App Store launching with 500 iPhone Applications, 25% Free," *Endgadget.com,* http://www.engadget.com/2008/07/10/jobs-app-store-launching-with-500-iphone-applications-25-free/ (accessed 14 March 2011).

17. Ibid.

18. Apple. 2009, November 4, "Apple Announces Over 100,000 Apps Now Available on the App Store" http://www.apple.com/pr/library/2009/11/04appstore.html (accessed 11 April 2013); Apple. 2011, July 7, "Apple's App Store Downloads Top 15 Billion" http://www.apple.com/pr/library/2011/07/07Apples-App-Store-Downloads-Top-15-Billion.html (accessed 13 November 2012).

19. http://cydia.saurik.com/ (accessed 22 December 2012).

20. http://itunes.apple.com/us/app/talking-tom-cat/id377194688?mt=8 (accessed 24 June 2013).

21. Don Ihde, *Technology and the Lifeworld: From Garden to Earth* (Bloomington and Indianapolis: Indiana University Press, 1990); Don Ihde, *Postphenomenology and Technoscience: The Peking University Lectures* (New York: State University of New York Press, 2009).

22. Ihde, *Technology and the Lifeworld*, 98; Don Ihde, *Heidegger's Technologies: Postphenomenological Perspectives* (New York: Fordham University Press, 2010), 48–49.

23. Cf. Ihde, *Technology and the Lifeworld*, 98–101. In the words of the *New York Times*, "anthropomorphism, like after-shave, is best used sparingly" (*New York Times* (Editorial), 2005, "Penguin Family Values," September 18, 2005, http://www.nytimes.com/2005/09/18/opinion/18sun2.html (accessed 5 April 2011)).

24. Ihde, *Technology and the Lifeworld*, 98. A major technological ingredient in the quasi-otherness of the cat and other virtual pet figures is the cell phone's screen that displays its face at a high-enough resolution. However, these are not real faces—they are quasi-faces of quasi-others (see Chapter 6).

25. Loechner, Jack, "Moms Mobile Addiction," *Center for Media Research*, April 14, 2011, http://www.mediapost.com/publications/?fa=Articles.showArticle&art_aid=148416 (accessed 27 April 2011).

26. Ibid.

27. Cf. Goggin, "Adapting the Mobile Phone," 232.

28. Leslie Haddon, "Domestication and Mobile Telephony," in *Machines That Become Us: The Social Context of Personal Communication Technology*, ed. James E. Katz (New Brunswick and London: Transaction Publishers, 2001), 46.

29. See Roger Silverstone, Eric Hirsch, and David Morley, "Information and Communciation Technologies and the Moral Economy of the Household," in *Consuming Technologies: Media and Information in Domestic Spaces*, ed. Roger Silverstone (London: Routledge, 1992), 18.

30. Ihde, *Technology and the Lifeworld*, 144–150.

31. Cf. Jason Farman, *Mobile Interface Theory: Embodied Space and Locative Media* (New York and London: Routledge, 2012).

32. Biggs, John, "Has The Age of Totemic Gadgets Passed?" *CrunchGear* April 2, 2011, http://www.crunchgear.com/2011/04/02/has-the-age-of-totemic-gadgets-passed/ (accessed 20 April 2011).

33. Carole Riviere, "Mobile Camera Phones: A New Form of 'Being Together' in Daily Interpersonal Communication," in *Mobile Communications: Re-negotiation of the Social Sphere*, ed. Rich Ling and Per E. Pederson (London: Springer, 2005), 176; Rouse Margaret, m-payment (mobile payment), *Search Mobile Computing*, September 2005, http://searchmobilecomputing.techtarget.com/definition/m-payment (accessed 8 November 2012); Wayne Cunningham, "The smart cell key phone," *CNET*, September 24, 2008, http://www.cnet.com/news/the-smart-cell-key-phone/ (accessed 23 May 2015).

Chapter 4

Fourth Historical Variation

Sensory Exploration[1]

HISTORIES OF THE FUTURE IN THE MAKING

So far I have described three historical variations of the cell phone that have already come about. But this is not sufficient. "Interpretation . . . across past temporal distance cannot remain the only direction for contemporary hermeneutics. It must also turn to the 'possible worlds' of the future."[2] The analysis of the cell phone should not stop at this point, and needs to be continued. This chapter is a "history of the future" of the cell phone; it considers the potential of a future fourth historical variation of the cell phone with an analysis that is grounded in today's technologies. A retrospective look at the first and second variations of cell phones shows that features of each one provided indications for the design and capabilities of future variations. For example, texting was theoretically available in the first variation to StarTAC users and later became a significant feature of cell phones of the second variation. Similarly, the presence of a few applications rendered the second variation a gateway for the mostly application-based third variation. The capabilities of the iPhone (and other third variation smartphones) that render these cell phones into devices that "sense" and decipher some aspects of their surroundings serve as the starting point of this chapter's analysis of the future-oriented fourth historical variation. As this historical variation is still forming, the possible trajectories its formation may take are numerous. In this chapter I assess one possible direction, namely augmented reality, based on functionalities that can be found in today's cell phones, which are collectively termed here as "sensory exploration."

When I originally wrote this chapter, this variation was taking in its earliest stages. Today, wearable computing is already hyped and sensory exploration has begun to materialize. Still, unlike the first three variations where I could depict a handset that dominated the discourse, I cannot point at a specific handset or device as the emblem of this variation: GoogleGlass is not a mass-market product; Fitbit and other bracelets that monitor human activity have limited capabilities; and smartwatches were only recently announced.[3] What is interesting about these wearable devices is their connection to the cell phone and the way they participate in the distribution of its capabilities along the human body. Also interesting is the new ways in which the wearable devices collect data, or in other words, present new types of input to the cell phone. Consequently, the balance between input and output mechanisms has been shifting. This balance is metastable and always moving, as can be evidenced throughout the first three historical variations. The first variation *outputed* the CallerID on a small LED screen; the second variation was focused on the *input* of text by the user; the third variation shifted back to *output* through a relatively large screen showing multimedia; and now the pendulum moves back to *input*, exceeding the keypad paradigm of the second variation.[4]

It should be noted that most of these input and output techniques are focused on seeing and hearing. Other senses are used in a limited way. For example, touch in the context of the cell phone is almost equivalent to vibration—mostly used for alerting but also imaginatively applied as a massage application.[5] Smart bracelets can analyze the skin's responses but they are confined to output (i.e., reading it), and there is no input on the part of the device (i.e., effect on the user). Smell and taste could have been possible directions for inspection, but I could not find significant signs for such output technologies based on chemical senses. These senses, along with the sense of touch, are sometimes termed "the intimate senses" to denote the intimate relations they require between bodies and environments. They can be a possible direction, but do not significantly manifest in previous historical variations. I leave it to history to decide if this should become the fifth or sixth historical variation. It remains to be seen when our smartphone, smartglasses or smart-X will be able to convey the taste of a dish as part of the information about a restaurant.

Various *input* technologies have been multiplying and developing since the third variation, and more extensively into the fourth variation. Some of them present surprising trajectories: the **keypad** evolved into keyboard for some smartphones, and the keyboard became virtual and part of a (touch) screen (for some devices), just to be bypassed by spoken commands (e.g., iPhone 4S's Siri or GoogleGlass oral commands); the **microphone** not only "hears" voice conversations, but can also be used to identify a song by its title and performer (e.g., the Shazam application);[6] a miniaturized electronic **compass** "knows" the direction in which the handset is pointed;[7] built-in **sensors** can notice how the handset is held (portrait or landscape, near the ear or in front of the face), current ambient light, altitude, or in the case of bracelets, sensors notice the user's metrics (e.g., movements made, heart rate, skin temperature, or galvanic skin response that indicates levels of stress); an embedded **camera** enables the cell phone to "see" with constantly growing accuracy and can track not only still images but also videos; and an integrated Global Positioning System (**GPS**) allows the cell phone to identify its location in space. These sub-technologies, parts of our embodiment and hermeneutics are delegated to the cell phone.

I begin this chapter with a review of the built-in camera and the GPS. The GPS is chosen to represent input technologies that rely on signals that are unperceivable by humans without the mediation of technologies; the camera represents input technologies that simulate a human sense. Both technologies were originally developed for independent devices and, through their integration into the handset, dramatically transformed the mediation of the world. This review may exemplify how the notion of histories of the future can become a useful tool for technological design, development and forecasting. In the next section, I examine the impacts of the fourth historical variation on perceptions, embodiment and hermeneutics.

BUILT-IN CAMERA

In an interview with *Wired* magazine, Apple's CEO Steve Jobs described the development process of technologies in terms of creativity:

> Creativity is just connecting things. When you ask creative people how they did something, they feel a little guilty because they didn't really do it, they just saw something. It seemed obvious to them after a while. That's because they were able to connect experiences they've had and synthesize new things.[8]

The development of a new product or feature is not about solving a problem or responding to a need. Instead, it consists of a playful connection of existing technologies (in Deleuzian terminology one could say that from the aggregate, an assemblage is born). This is probably how the cell phone got its built-in camera.

The integrated cell phone camera was first introduced in Japan in 1999.[9] In the first years, the uses of the cell phone-camera combo remained in the realm of taking pictures and then sharing them by either displaying them on the cell phone's screen, sending them by e-mail or posting them on the Internet.[10] Cell phone researcher Barbara Scifo identifies "two different levels of experience: on an individual level, the camera-phone represents a new form of extension of one's experience and memory, while on the socialization level, it represents a new communication resource that can be invested in one's peer group and in one's relationships."[11] These two levels were later developed in the literature. The first was described as "capturing the 'fleeting and unexpected' in the everyday,"[12] or more generally as "personal news gathering"[13] in that it includes the expected and the unexpected alike. The two levels were merged in the notion of "collective archive."[14] The two levels could be merged because the camera kept its original role as a visual recorder, whether as a memory extension or a form of communication.

A significant development occurred when some mobile applications started using the camera for more than taking pictures. They utilized the cell phone-camera combo to extend the hermeneutic capabilities of the user. For instance, the camera "reads" a barcode and the cell phone presents the lowest price of the product that carries the barcode in nearby shops; the camera "reads" a word and the cell phone translates it; the camera "reads" a business card and the cell phone stores the contact details in the address book.[15] These applications transfer the hermeneutic work from the human to the cell phone when the language is foreign, as in the case of translations, or when machine reading is needed, as in the case of barcodes and Quick Response (QR) codes that humans cannot read.

In parallel with these point-and-decipher applications, an alternative mode emerged in which applications help the user refer to the environment in a richer way. This mode is known as "augmented reality" or "AR." The basic function of these applications allows users to view reality through the camera's lens and then displaying it on the cell phone's screen. On top of this basic view, the cell phone presents a layer of information. Lev Manovich defines AR as an application that "dynamically deliver[s] dynamic data, or extract[s] data from, physical space."[16] The physical space is augmented by an "augmented space" or a "cellspace."[17] This space integrates data from the Internet with data from objects located near the user, so that the user obtains additional knowledge of the world. One possible way of integrating physical and augmented spaces is offered by an iPhone application named "Time Shutter San Francisco."[18] It asks the user to take a picture of an old building in the historical district of San Francisco, and then presents that building with an overlay of a vintage photograph of that place. I classify this application as AR because the additional information it provides is "directly related to the user's immediate physical space."[19] The Time Shutter application does not present the whole history, but only a segment of it that matches the user's current location. One can think of how additional filters beyond location, such as the user's preferences and interests, can result in further personalization of data.[20]

When I updated this chapter, I wanted to add a reference to the AR capabilities of GoogleGlass that are said to present layers of information while the user looks through the eyeglasses. That was indeed the original intention of GoogleGlass' developers as can be seen in an early concept video clip.[21]

At the time of writing, the presentation of map directions, for example, is on GoogleGlass' small screen on the upper corner of the right eye, which only allows the user to focus the eyesight on either the real world or on that screen (it should be noted that this mode-of-operation of GoogleGlass does not use the built-in camera). The way to get both the real and the virtual simultaneously is via the vocal direction commands, which are heard as the user moves. The visual layers, as these words are being written, are (still) separated.

To sum up, the contributions of the built-in camera to the cell phone experience can be mapped into two types: one is the provisioning of a pictorial background on which the layers of information can be presented; the other is the decipherment of selected visual signs. The first is an "augmented space," the second can be termed a "hermeneutic extension." One is about presentation and embodiment, the other is about hermeneutics and exploration. The augmented space—and to some degree the hermeneutic extension—can present more relevant data if the cell phone "knows" where it is. The San Francisco tourist application needs to "know" where the user is in order to present the right vintage picture. This is where the cell phone-camera combo (as well as wearable devices) is supplemented with GPS capabilities.

BUILT-IN GPS

The globe is covered by satellite positioning pulses that allow a GPS module of a cell phone to accurately determine its current location. In May 2000, the US government removed the encryption of the pulses, thereby opening its military GPS to civilians.[22] In 2003 some cell phones integrated a GPS module,[23] but it was the launch of a GPS-enabled iPhone in 2008 that marked the mass usage of cellular GPS navigation.[24] Recently, the GPS component has also been used outside the cell phone and not necessarily connected to it, as in the case of integrated cars' GPS, wearables or pets' collars. In such cases, these devices may still need to connect with a cell phone so that the cell phone transmit information to it from the Internet (e.g., traffic updates), display information (e.g., a map showing the running route), or store the data produced by the GPS-carrying device. In all of these examples, a relevant application must be installed on the cell phone.

Cell phone applications that use the GPS capabilities are known as "location-based services" (LBS). At a basic level, such applications notify users of the location of their cell phones (or the external device connected to it, usually wirelessly). These notifications can be external; for example, they can be sent to emergency services or to a user who lost his or her cell phone. They can also be used internally, as in the case of adding location details to a picture taken via the embedded camera. In case the user moves, a set of locations is used by the cell phone to calculate distance and speed, so that the cell phone can serve as a personal trainer for running and other sport activities.

Many of these applications function like a map on which the user's coordinates are marked, either in the singular as in the case of the location where a picture is taken, or in the plural as in the case of running and walking routes. Before the GPS and LBS were developed, maps and routes were conceived separately. Michel De Certeau distinguishes between maps and mere movement in space, such as walking. He writes,

> It is true that the operations of walking on can be traced on city maps in such a way as to transcribe their paths (here well-trodden, there very faint) and their trajectories (going this way and not that). But these thick or thin curves only refer, like words, to the absence of what has passed by. Surveys of routes miss what was: the act itself of passing by.[25]

De Certeau bundles maps, city plans and overviews from the top of buildings as theory, and juxtaposes it with walking trajectories, which he considers a practice. One is static, the other is dynamic. One is clean, clear and cold, the other is dirty, crowded and hot. One freezes reality and therefore creates a past, the other refuses to record the passing-by, the flux, and is built "on the fly." One is always-already outdated, the other is up-to-date up-to-the-minute. I suggest that the new navigation mode enabled by a cell phone equipped with a GPS module and LBS applications combines these extremes and softens the dichotomy. In other words, Manovich's notion of cellspace seems to merge De Certeau's theoretical with the practical, the predetermined past with the unstable present. The merger occurs when certain LBS applications integrate city plans and dynamic data from the Internet such as people's locations. Together they create augmented reality.

Basically, augmented reality applications provide supplementary information that is relevant to the current location, like tourist attractions, restaurants or ATMs in the vicinity of the location detected by the GPS.[26] However, information is not limited to data related to objects on the ground. It can also refer to the sky. For instance, "Starmap" is an application that provides a map of stars, their names and their groups as an overlay on an image of the portion of the sky the user is pointing at with their device.[27] Starmap presents only the part of the sky that is visible at the specific geographic location the user is located in and the part of the sky the user is pointing at. The information provided on the cellspace does not have to be knowledge; it can be imaginary, as in the case of the fitness iPhone application "Zombies, Run!" in which the user runs and hears voice instructions as if he or she is escaping from zombies.[28] In this category, one may find games that use the GPS functionality to connect between the real world's cityscapes and the virtual space of a game.

Passing through a certain physical location entitles the user/player to virtual points. Sometimes these games include social awareness, so that friends' locations are marked.[29] And sometimes, reversely, the social network itself is the focus through members' sharing of their location details, as in the case of Foursquare, an application through which users share their location with friends by "checking in."[30] These examples demonstrate how flexible the information for each specific location can be. It can be static, like touristic information, or dynamic, like to-the-minute updates about friends who are nearby. Like in quantum physics, the mere presence of the user at a certain spot changes the information associated with that spot. In postphenomenological terms, the hermeneutic relations that a user maintains with the cell phone with regard to reading the surroundings are not constant. The information may change any minute and one cannot expect to receive the same data on multiple visits or "check-ins" at a certain spot.

This last example of Foursquare matches the term "hermeneutic extension" that I developed for the built-in camera. The other examples fit into the "augmented space." In the next section I will further elaborate on the changes in embodiment and hermeneutic relations of the fourth historical variation. But before that, I will discuss the changes in perception brought by this variation.

THE IMPACTS OF THE FOURTH HISTORICAL VARIATION

The Mechanization of Perception

The input technologies associated with the microphone, sensors, camera and GPS enable the cell phone to decipher the user's environment. Other input technologies decipher the user's own body, especially when built into "wearables" such as bracelets and smart watches. Through these input technologies, the cell phone knows the environment in a way similar to the "immediate and focused bodily [impression] in actual seeing, hearing etc.,"[31] which Ihde calls a micro-perception. A micro-perception is sensory in nature and is complemented by the macro-perception that is "the hermeneutic-cultural signifying dimension of a situated experience."[32] The macro-perception facilitates a common experience to be shared by several individuals with a similar cultural background.[33] Different micro- or macro-perceptions create different experiences. Micro-perceptions can be common when they are delegated to technologies. A cell phone of the fourth variation can micro-perceive the environment: with an integrated camera it "sees" the fragments of the reality to which the user points; with the microphone it "hears" music and recognizes it;[34] and with a GPS it "knows" where it is. Complementarily, wearable devices micro-perceive the user's body: how many steps were made, heart rate, skin parameters (to detect stress levels), etc.

Some technologies attempt to produce macro-perceptions, such as associating the same meaning to a certain heartbeat, or assuming sleep when a certain motion pattern is detected. Being constantly connected to the Internet may enable the provisioning of a more nuanced and flexible social context to certain micro-perceptions, thereby enriching the macro-perceptions.

Certain types of information provided by the cell phone and its related devices cannot be micro-perceived without technological mediation. Knowing one's own heartbeat is difficult to achieve on a constant basis without the technological mediation of a smartwatch or a monitoring bracelet. My next example has a richer perceptual input, yet it involves a standard operation like walking down the street. Traditionally, we see various signs that help us decipher what is around us. Augmented reality unveils additional information that is hidden, like the existence of a small business operating in a backyard with no sign in the front, a toilet in a basement of a nearby store, or other people's critiques of a restaurant. Such physical sites are "geo-tagged" (so they can be detected by the relevant application), the landscape is transformed into more than a physical environment and the "physical" micro-perceptions are supplemented by the augmented space.[35] From the user's perspective, the additional information de-centers the human sight from the primacy it had. As a result, the hermeneutic relations we have with our environment become increasingly dependent upon the cell phone's technological mediation. Even simple navigation such as driving to work and back home becomes dependent on technological mediation in order to find a route that is less congested. Though such a route is well known, technological mediation is necessary to calculate the changing traffic conditions and suggest alternative routes.

These new dimensions of reality come with a price, because they are produced through the delegation of some perceptions to a technological artifact, in this case the cell phone. We have to trust these artifacts, their application developers and those involved in their operations. In return, we have a cell phone that "knows" our surroundings in ways we could not achieve alone.

The Cellular City

As AR starts taking off, some tangible changes in the environment are already happening. The world which is being augmented physically changes by the addition of QR codes, that is—a version of barcodes that can be captured by cell phones' built-in cameras (provided the suitable application is installed). These codes become part of street advertisement, so that people can point their cell phone's camera to a corner of an ad where the QR code appears and receive videos, music and additional information on the product or service advertised.

Jason Farman describes an interesting example of the façade of the N Building in Tokyo, which is a large QR code. People passing by the building can point their cell phone's cameras to the façade and get information and updates on the businesses and people in the building. Without a camera connected to a smart device, may it be a cell phone or glasses, the building's façade is meaningless and the associated information is mostly inaccessible.

New Embodiment, New Hermeneutics

When perceptions are delegated to a technological artifact like the cell phone, the bodily comportment is transformed and new embodiment relations are established. One of the classical examples of embodiment relations is the blind man's cane that transforms not only the man's bodily movements but also mediates the world for him and supplements his missing vision. Another classical example is eyeglasses with which the world is more visible and clear. Similarly, the cell phone renders the landscape into more than a physical environment, that is, it makes the landscape more intelligible through the new embodiment (and hermeneutic) relations it offers. It is interesting to note that the postphenomenological examples for transformative embodiment relations (the cane and the eyeglasses) are taken from the terminology of blindness and poor eyesight. AR may become our new metaphorical cane and eyeglasses, but instead of improving a defective organ, these technologies add new capabilities as well as new experiences. The smart phone and the smart glasses do not "see" instead of us, but rather add a new dimension to our standard eyesight. In this sense, the telescope is a more suitable analogy because it shapes not only a new embodiment but also a new hermeneutic. Let us refine the analogy to the telescope and consider the technologies of the fourth historical variation as operating like X-ray or radio telescopes that read the invisible electromagnetic radiation of a hidden star. Likewise, AR technologies add capabilities that are not naturally given. They are not a prosthesis that replaces what is missing or defective. They do not aim at rendering our bodies perfect. What they do is add new viewpoints to our bodily experience, modify the hermeneutic relations we have with the world around us, and produce new types of information about us and about the world around us. It would be interesting to think of a non-cell phone non-eyeglass artifact that produces AR, perhaps in the form of a cane or a telescope.

Through new embodiment and hermeneutic relations, the fourth historical variation offers a new mobility experience. It is convenient (and even addictive) to open a navigation application when starting to move. It informs us when the bus is coming, alerts on traffic congestions and may offer a coupon for a cup of coffee in the newly opened coffee shop on the way. For Adriana de Souza e Silva and Jordan Frith, this means that information "becomes part of that space, not just an augmentation to it."[36] As information becomes part of a space, that space turns into a place. Now it is a place in the world that is enriched with additional information that is accessible at a certain location, thereby enhancing the attachment of the virtual to the real. However, such an analysis does not explain why people change their behavior and invest time and money in these new technologies. The user's perspective offers an alternative explanation. It consists of the conceptualization of the relations between the two spaces as the matching of two different contexts—the world around the user and the additional information. AR technologies mediate these two contexts and match them through embodiment and hermeneutic relations, and thereby turn the environment into a more intelligible space. When we use our cell phone we know why the bus is late; we know more about the stars; we know more about the restaurants near our hotel in a town we have never visited before. Moreover, the additional information produces a different experience for each user because different details are revealed to various users according to their profile and preferences (i.e., through personalization), or according to the time of day and other filters. This change is both epistemic and sensory; in postphenomenological terms, the fourth variation offers a new bundle of embodiment and hermeneutic relations with the technology that transforms us and our world.

New Experiences of Mobility

Technologies of the fourth variation not only change existing movements but also add new movement possibilities. At this point, I find very useful the analysis made by de Souza e Silva in which she examines the mobility element of location-based social networks (e.g., Foursquare) with Deleuze and Guattari's theoretical frameworks of smooth and striated spaces. Simply explained, a striated space is one in which movement is predetermined, as it is in the case of streets and lanes in which one is obliged to move in certain directions. In contrast, the smooth space allows movement in any direction, as in the case of the sea or the desert where 360-degree mobility is possible. In the context of GPS navigation, these frameworks are useful for conceptualizing the impact of the technology on its users and the world. I refer here to the new paths that navigation software enables through the recording of people's movements. If a user knows a shortcut that is not documented on the maps, the mere movement within this shortcut creates a sign that other users may follow. If more people use this shortcut, then it becomes more dominant among the recommended paths the system offers. It is a "smoothing" of the striated cityscape. The opposite occurs when users cross a park via a new path in the middle of a forest or a lawn, so that the smooth becomes striated. This bidirectional effect leads to the conclusion that the new mobility enabled by the cell phone cannot be understood to merely "quicken . . . the pace of urban life and at an aggregate level, resulting in a dramatic increase in the metabolism of urban systems promote urban lifestyle."[37] In addition to studying how the cell phone promotes urban lifestyle, there should be an analysis of the changes introduced into the suburban lifestyle, as well as that of the countryside.

The new meanings ascribed to existing movements combined with new possibilities for movements point to the importance of movement in AR applications. Unlike virtual reality applications, which require a sedentary bodily position, AR not only allows movement in the world, but even calls for it. AR applications dictate active and dynamic embodiment relations. Wearable devices support and contribute to these kinds of embodiment relations. Specifically, wearables extend the input and output mechanisms beyond the cell phone and distribute them along the user's body, and therefore indirectly support movement in the world. For instance, wearables that monitor the user's body presuppose a body that is in the world, immersed in reality and changing according to the environmental conditions. In this way, AR can be regarded as a backlash against virtual reality. AR does not aim at producing an alternative reality but instead attempts to augment what we have with additional information.

These new experiences of mobility enabled by AR bring the ratchet effect into play, which reminds us how difficult it is to go backward to a previous technological state. Like someone who got used to wearing eyeglasses and now cannot view the world without them, so too the traditional pure reality worldview may seem shallow and insufficient once AR's ocean of information unfolds through the cell phone's screen, the smartwatch or certain smart bracelets.

In this chapter I reviewed the already visible signs of the fourth historical variation. Although there is not (yet) a handset (or another device) that is designed to fully microperceive the environment, I have tried to hint at how perceptions may be modified in such a technological landscape with the postphenomenological notions of embodiment and hermeneutic relations and micro- and macro-perceptions. This fourth historical variation is an attempt to write histories of the future, thereby answering the postphenomenological challenge of "how does one do phenomenology, rather than discourse about it or engage in internecine interpretations."[38]

NOTES

1. A version of this chapter was published as an article "No Longer a Phone: The Cell Phone as an Enabler of Augmented Reality," *Transfers: Interdisciplinary Journal of Mobility Studies* 3.2 (2013): 70–88.

2. Don Ihde, *Expanding Hermeneutics: Visualism in Science* (Evanston: Northwestern University Press, 1998), 25.

3. When Apple launched its smartwatch it was already defined by the critique as "future ennui," see Ian Bogost, "Future Ennui," *The Atlantic* 16 September 2014, http://www.theatlantic.com/technology/archive/2014/09/future-ennui/380099/ (accessed 23 May 2015).

4. Adding a miniature screen on a smart watch in the fourth variation is in line with attempts to enlarge and extend the cell phone's screen in its third historical variation. All of these screens are based on similar technologies. For instance, there is no holographic display on these screens, which can significantly transform the experience.

5. https://itunes.apple.com/us/app/vibrating-massager-free/id312272622?mt=8 (accessed 13 November 2012).

6. http://www.shazam.com/ (accessed 27 January 2011).

7. It is useful for religious Muslims who need to know the direction of Mecca, see "A cell phone that points to Mecca," *MSNBC* 10 July 2005, http://www.msnbc.msn.com/id/9622362/ns/technology_and_science-wireless/t/cell-phone-points-mecca/#.UCOrC2FgIwM (accessed 31 July 2012). This is a typical example of a functionality that existed well before the iPhone, but the context of sensory understanding gives it new directions.

8. Gary Wolf, 1995, "Steve Jobs: The Next Insanely Great Thing," *Wired* February 1995, http://www.wired.com/wired/archive/4.02/jobs_pr.html (accessed 8 November, 2012).

9. Gerard Goggin, *Cell Phone Culture: Mobile Technology in Everyday Life* (London and New York: Routledge, 2006), 144.

10. See Daisuke Okabe, "Emergent Social Practices, Situations and Relations through Everyday Camera Phone Use," *Mobile Communication and Social Change* (Seoul, Korea, 2004); Carole Riviere, "Mobile Camera Phones: A New Form of 'Being Together' in Daily Interpersonal Communication," in *Mobile Communications: Re-negotiation of the Social Sphere*, ed. Rich Ling and Per E. Pederson (London: Springer, 2005), 167–185; Jane Vincent, Leslie Haddon, and Lynne Hamill, "The Influence of Mobile Phone Users on the Design of 3G Products and Services," *The Journal of the Communications Network* 4, no. 4 (October–December 2005): 69–73; Barbara Scifo, "The Sociocultural Forms of Mobile Personal Photographs in a Cross-Media Ecology: Reflections Starting from the Young Italian Experience," *Knowledge, Technology and Policy* 22, no. 3 (2009): 185–194. Many users ignored the possibility of sending pictures over the cellular network as MMS (Multimedia messaging Service), which is a text message with multimedia content. The frequently mentioned reasons were the relatively high cost compared to a text message and complicated operation.

11. Barbara Scifo, "The Domestication of Camera-Phone and MMS Communication: The Early Experiences of Young Italians," in *A Sense of Place: The Global and the Local in Mobile Communication*, ed. K. Nyiri (Vienna: Passagen Verlag, 2003), 365.

12. Harvey May, and Greg Hearn, "The Mobile Phone as Media," *International Journal of Cultural Studies* 8, no. 2 (2005), 205.

13. Goggin, *Cell Phone Culture*.

14. Nicola Green, "Mobility, Memory and Identity," in *Mobile Technologies: From Telecommunications to Media*, ed. Gerard Goggin and Larissa Hjorth (New York and London: Routledge, 2009), 266–281.

15. Examples of such already existing applications: "Barcode Reader," http://itunes.apple.com/us/app/barcode-reader/id340825499?mt=8; "Babelshot (photo translator)" http://itunes.apple.com/us/app/babelshot-photo-translator/id334194705?mt=8; "Business Card Reader" http://itunes.apple.com/il/app/business-card-reader/id328175747?mt=8 (accessed 6 April 2011).

16. Lev Manovich, "The Poetics of Augmented Space," *Visual Communications* 5, no. 2 (2006), 221.

17. Manovich, "Augmented Space," 220, 221. Adriana de Souza e Silva in her article "From Cyber to Hybrid: Mobile Technologies as Interfaces of Hybrid Spaces," *Space and Culture* 9, no. 3 (August 2006): 261–278 rejects the notion of augmented space and suggests instead "hybrid space." I prefer Manovich's conception of augmented space because he defines how the digital and the physical (or virtual and real) interact, while "hybrid space" is loosely defined as "blurring the borders between digital and physical spaces" (263). De Souza e Silva further maintains that these borders were "apparently clear" when Internet was stationary and used at homes and offices (264). It seems to me that even for stationary Internet, these borders have never been clear. Take, for example, Amazon.com which has existed since 1995 (the days of the first historical variation)—is it only a virtual shop or a shopping enabler and a substitute for "physical" shops? Therefore, the blurred line between the digital and the physical is not unique to the cellspace and cannot serve as a differentiator. Another element in De Souza e Silva's definition of hybrid space is its sociality. The hybrid space, she claims, is created "because of the use of mobile technologies as social devices" (265). As I show here, the use of cell phone is not limited to the interaction between two people, but can also be for additional goals, such as entertainment, memorization and education. Additionally, I find it difficult to refer to a cell phone as "a social device." The term social, following Bruno Latour, should be the explanandum (the phenomenon to be explained) and not the explanan (the explanation of that phenomenon), the consequence and not the cause. Yes, the cell phone has social effects, but its repertoire of effects is wider, as I attempt to show in this book. Moreover, De Souza e Silva stresses that "a hybrid space is *not* constructed by technology. It is built by the connection of mobility and communication and materialized by social networks" (265). In postphenomenological terms, the technological cannot be divorced from any space. Employing the I-technology-world formula in one's analysis makes it impossible to determine that a certain structure is not mediated by any technology. Wherever and whenever humans are involved, there is a technology that mediates the world, directly or indirectly. Therefore, in this book I remain faithful to Manovich's definition of augmented reality as the layering of the virtual over the real, and do not use hybrid or mixed space.

18. http://itunes.apple.com/il/app/time-shutter-san-francisco/id411557094?mt=8, accessed 27 January 2011.

19. Manovich, "Augmented Space," 225.

20. Manovich warns of the close connection between surveillance and augmented reality, because the provisioning of information that matches location, preferences and mood requires previous monitoring of that user to learn his or her taste.

21. Nick Bilton, "Google Begins Testing Its Augmented-Reality Glasses," *The New York Times*, April 4, 2012 (http://bits.blogs.nytimes.com/2012/04/04/google-begins-testing-its-augmented-reality-glasses/, accessed 1 September 2014). The video clip appears at the end of the article.

22. Rick W. Sturdevant, "NAVSTAR, the Global Positioning System: A Sampling of Its Military, Civil, and Commercial Impact," in *Societal Impact of Spaceflight*, ed. Steven J. Dick and Roger D. Launius (Government Printing Office, 2009), 331–351 (http://history.nasa.gov/sp4801-chapter17.pdf, accessed 24 June 2013), 332.

23. Motorola had GPS-enabled cell phones since 2003 (e.g., A835, A920 and A925). However, without the proper software applications, the GPS was useless. In Internet forums an E1000 user named Bastian asked in November 2004: "Is it true that the E1000 supports a GPS module?" (http://www.tomshardware.com/forum/42812–38-motorola-e1000-modul, accessed 1 December 2012). This model was launched at the beginning of 2004 and the local cellular operator was in charge of supplying the relevant software. Bastian received a reply within few days: "I know that at least one Company here is working on a solution that could make a GPS useful. (Yes, I am involved in that development)" (ibid).

24. Adriana de Souza e Silva and Jordan Frith, "Locative Mobile Social Networks: Mapping Communication and Location in Urban Spaces," *Mobilities* 5, no. 4 (November 2010): 485–506.

25. Michel De Certeau, *The Practice of Everyday Life,* trans. by Steven Rendall (Berkeley and Los Angeles: University of California Press, 1984), 97.

26. Goggin, *Cell Phone Culture*, 196–7. Within this category, Manovich also mentions in his article "Augmented Space" checking-in at the airport or paying for a road toll, which can be activated only at specific locations even though GPS is not necessarily used.

27. http://itunes.apple.com/us/app/starmap/id284408099?mt=8# (accessed 6 April 2011).

28. https://www.zombiesrungame.com/ (accessed 7 January 2013).

29. Claire Dormann and Robert Biddle. "A Geo-Wiki Game for Urban Exploration and Community Participation," Workshop: Exploring the Practices & Technologies of Representing the Urban Environment (San Jose California, 2007).

30. https://foursquare.com/ (accessed 13 November 2012). Checking-in is a notification sent to the social network via a smart phone app or by text message. Additionally, the application has a game element: points are awarded for checking-in at various venues. By checking-in a certain number of times, or in different locations, users can collect virtual badges. In addition, users who have checked-in most of the times at a certain venue will be crowned "Mayor" until someone surpasses their number. Some venues offer special deals to users who are "mayors" (http://www.crunchbase.com/

company/foursquare, accessed 13 November 2012). Other applications with similar functionality were reviewed by de Souza e Silva and Frith in "Locative Mobile Social Networks," but most of the companies that developed these applications no longer exist as of 2012. One explanation may be users' reluctance to automatically share their locations. Unlike Twitter, in which the user selects when to share a thought, the locative mobile social networks disseminated the user's whereabouts whenever the user was logged in. That was considered by many as a risk to their privacy, as de Souza e Silva and Frith rightfully warn in their paper.

31. Don Ihde, *Technology and the Lifeworld: From Garden to Earth* (Bloomington and Indianapolis: Indiana University Press, 1990), 29.

32. Don Ihde, "A Response to My Critics," *Techné* 7 (2003): 114.

33. It also explains the different experiences that individuals from divergent cultures have in regard to the same technology. There is no imaginary, single, unified true-for-all macro-perception associated with a specific technology, and the notions of multistability and pluriculture form the caveat for such wrong assumptions, Peter-Paul Verbeek, *What Things Do: Philosophical Reflections on Technology, Agency and Design*, (University Park: The Pennsylvania State University Press, 2005), 135–138. For Ihde, the roots of macro-perception can be detected in Heidegger's thoughts, see *Consequences of Phenomenology* (Albany: State University of New York Press, 1986), 58.

34. For example, Shazam (see fn. 6).

35. See Jason Farman, "Mobile Media Performances as Asynchronous Embodiment," *The International Hournal of Screendance* 2 (Spring 2012): 48–51.

36. de Souza e Silva and Frith, "Locative Mobile Social Networks," 490.

37. Anthony M. Townsend, "Life in the Realtime City: Mobile Telephones and Urban Metabolism," *Journal of Urban Technology* 7, no. 2 (2000), 85. Cf. de Souza e Silva, "From Cyber to Hybrid."

38. Don Ihde, *Postphenomenology - Again?* Working Papers from Centre for STS Studies, University of Aarhus, The Centre for STS Studies, 2003.

Interlude

Transforming "I," "Technology" and "World"

In the last four chapters, I have characterized four historical variations of the cell phone: the first is focused on voice communications; the second's defining features comprise short texts, either as messages or as notes; for the third, applications and large screens are conflated; and the fourth orients toward sensory exploration of the environment through various input technologies. This description of the cell phone's evolution demonstrates the various meanings it carries for various users at various times. It also reveals why it is so difficult to determine when it became more than a mere phone.

In order to trace the specifically *technological* evolution of cell phone technology, I selected a particular cell phone model as the protagonist of each historical variation. Yet, the analysis in these chapters not only described modifications in the technology itself, but also modifications in the other two components of the I-technology-world scheme. The next two subsections will summarize the changes in the "I" and in the "world." The changes in the user's sensation of reality are revealed once the transformations in the "I" are accumulated, and parallel clarity is gained for the changes in one's experience of one's surroundings through the accumulation of changes in the world brought about by each of the four variations.

TRANSFORMATIONS OF THE "I"

In this subsection, transformations of the "I" are classified by the four postphenomenological relations of embodiment, hermeneutic, alterity and background.

Embodiment relations have evolved around the ways in which the cell phone occupies the hands and the gaze. In the first variation, the cell phone is held only during calls and gazed into for just a short while in order to see who is calling. In the second variation, the hold and the gaze are slightly longer because there are more texts to be read. The applications of the third variation require more time spent on gazing at and holding the handset, and in the fourth variation it is necessary to hold and look at the cell phone's screen most of the time in order to experience the world with the additional layer(s) of augmented reality. With wearable devices, communication is "always on," that is—the technological device participates in each and every moment of life. This evolution in embodiment relations demonstrates how the visual has taken over the auditory and has regained its supremacy as the major mode of sensing the world.[1]

Changes in hermeneutic relations occur as novel forms of writing appear in the first historical variation, and are formalized in the second variation. The cell phone has opened up a condensed format of writing, accompanied by new words in the dictionary (e.g., LOL in the *Oxford Dictionary*) and new signs (i.e., emoticons). This inventive hermeneutics has led to new formats for knowledge transfer (e.g., Twitter University). The new hermeneutic relations can be understood as being organized around the ways in which users construct their world. The cell phone participated in blurring the boundaries between home and work as calls, text messages and e-mails became available anytime, anywhere. It has also subverted the feeling of loneliness into what I called "the never-alone effect."[2] Children of working mothers are not alone at home since their mothers can contact them as part of their "remote mothering." Not only has the feeling of loneliness been diminished, but also the social category of being late. Micro- and hyper-coordinations through text messaging allow people to recalculate meeting time and lessen the feeling of being late. Moreover, the exact starting time of a meeting can be adjusted in real time with the assistance of navigation applications that make it possible to determine an estimated time of arrival. Alternatively, social network applications enable spotting nearby friends so that a meeting can be arranged almost spontaneously. Such new spontaneity is organized and intended, yet allows flexibility and may lead to a certain sensation of freedom from the abiding rules of timekeeping. Like the erosion in the feeling of loneliness and the loosening of time, getting lost loses its meaning when one is equipped with a smartphone or smart eyeglasses with a map application. Getting lost becomes a willful act and requires a decision to get lost, because the user can identify herself on the map at any minute. The meanings of being alone, being on time and getting lost have been melting away.

Thirdly, alterity relations have developed as the cell phone became an assistant involved in more and more areas of life. The device has been regarded as a companion that escorts the user and provides her or him with connections to others, to the world and to information as required. It started with contacts' phone numbers; developed into calendar, notes and extended contact information; expanded to multimedia; and as of the last variation, it can recommend a restaurant or a route through the filtering of information according to the user's location. In the third variation, the entertainment aspect of cell phone use has grown and expanded so much so that the cell phone functions not only as a butler and an instructor, but also as a clown. This companion is well adapted to the user's specific history, tastes and preferences through personalization and customization. No wonder, then, that alterity relations have evolved into dependency. Yet, the companion cannot be fully anthropomorphized. As the cell phone communicates via sound, sight and vibration, it leaves outside the full haptic range[3] and intimate senses of taste and smell.

Lastly, background relations have been transformed so that many technologies are reduced to applications—a telephone, a music player, a birthday card, a flashlight, and a television are all applications. The application paradigm of iPhone has taken over. This reduction to an application allows one to possess and hold lots of everyday-carry-equivalents, more than one could physically carry in one's pockets and in a bag, while still knowing that these carry-equivalents are there, just waiting to be operated. The address book of the first variation is no longer a paper booklet. Likewise, the notepad and the calendar are no longer in paper form and the accompanying pen or pencil have become redundant since the second variation. Coins for the parking meter are not necessary if you have a metro parking application (to be used in metered on-street parking areas). The reduction to an application has been even more acute in the fourth variation; while the functionality of the cell phone's camera and the compass has been increasingly enhanced, their sizes have been reduced to a chip. Along with the growth in the number of smartphone applications, the home as a space that is waiting for us in the background has been extended through the process of domestication that was described in the third variation. The home is no longer the single repository of photo albums or music records. Some of these are stored on the cell phone, and some in a cloud which can be accessed anytime anywhere via the cell phone or associated technologies, such as the tablet.

TRANSFORMATIONS OF THE "WORLD"

The cell phone has transformed the mix of our everyday carry to the extent that its collection of applications has practically replaced artifacts that used to populate the world. The transformation from artifacts to applications is eventually a change in the world. As the cell phone turned into the new Swiss army knife, many personal carry items were virtualized: writing devices such as pens became redundant; the watch turned into nothing more than a fashionable object; and house or car keys were replaced by yet another application. Sometimes the cell phone contributes to a virtualization process that started before its emergence. This is the case of money that in the digital age has turned virtual. With the cell phone, one is not required to carry even the last physical trace of money—the credit card.

Just as the mix of items considered as everyday carry is unique to each person, so is the combination of applications unique to each cell phone. The technological tendency of the cell phone, especially models since the iPhone, calls upon users to alter their phones according to their specific tastes and needs—an aspect that intensifies the multistability of individual objects. The multiplicity of combinations of applications results in an intensively multistable artifact. In terms of experience, hyper-personalization occurs when each user possesses a unique artifact, although all artifacts were identically manufactured.

The transformation of everyday carry from artifacts to applications led to changes in industrial ecosystems, such as the paper industry, in a process that began well before the emergence of the cell phone. It started in mid-1970s as "the paperless office" vision.[4] Paradoxically, within a few years offices turned into paper-intensive facilities. Meanwhile, discussions of the importance of becoming paperless continued as part of the sustainable development discourse. The cell phone has offered a new direction for a paperless world that starts in small steps. It replaces notebooks and memo pads with a virtual writing space, it provides cellular versions of parking payment slips,[5] and it delivers concert tickets and boarding passes in the form of an SMS/MMS containing a barcode. As books have become paperless, the cell phone has also participated in the growing e-book trend alongside e-readers. Recently, a headline in *The Atlantic* announced: "In China, 25 Million People Use Only Their Cell Phones to Read Books."[6] It seems that in China the book as a personal carry-on went virtual. Interestingly, this way of consuming literature leads many Chinese readers to prefer relatively short novels and poems that can be easily read on a small screen, according to the article. Preferences for shorter texts are in line with inclinations to condense knowledge, as was demonstrated by Twitter University and the introduction of emoticons to everyday discourse.

Another industry that has changed is the camera industry. With the cell phone, the camera is no more than a chip. While physical film disappeared more than a decade ago, boxes of cameras, now digital, remained relevant during the first years of the twenty-first century. Now they have also vanished, and those that remain are used almost exclusively by professional and semiprofessional photographers. Everyone else uses the cell phone's camera.

Like the everyday carry that became obsolete, some other artifacts that used to populate the public sphere have disappeared. One of the earliest disappearances was that of public phones, as described in the chapter on the first historical variation of cell phones. Instead, the public sphere is now inhabited by artifacts and technologies that are related to the operation of cell phones, such as cellular antennae, signs stating a place is Wi-Fi enabled, and QR codes. Others try to "hedge" the widespread use of cell phones by creating nearly cell phone-free spaces, such as train cars that are marked as "quiet zones." The emergence of QR codes in the public sphere is of special interest because they are intended to be seen and deciphered by cell phones. They are based on the delegation of micro-perceptions to an artifact and result in increased dependency on technological mediation.

The changes in the world are not only in the objects but also in their hermeneutics. Two examples demonstrate this change: translation and navigation. A cell phone of the fourth variation enables translation from one language to another. In *The Hitchhiker's Guide to the Galaxy,* Douglas Adams describes the Babel fish, a fictitious animal that instantly translates from any language to any language. The cell phone did not invent computerized translation, of course. But the ability to access Internet-based translation technology such as Google Translate application renders the cell phone an almost-Babel fish. The barrier of language is being eroded. Carrying a cell phone with a translation application is more than carrying a dictionary; it creates an effect of embodiment of unspoken languages.

Augmented reality applications also lead to a change in the movement of people. Modernity has praised the autonomous subject equipped with free will, whose movement is the result of a rational act. Under this assumption, traffic jams are the result of many independent decisions of people who want to get to work or back home. Navigation software that is used by large enough number of people may lead to a slightly different pattern of movement, that of a swarm, in which the actions of the individuals are coordinated. Every individual is informed about the state of the swarm and the application may advise several alternative routes. If the advice is accepted by enough users, eventually the state of the entire swarm is ameliorated as the traffic jam becomes less congested.

I have attempted to describe here how the cell phone has transformed the world on two levels: the "micro" level, which is close to the user, and the "macro" level, which is that of the world. Both types of changes did not occur abruptly. An overview of the historical variations reveals the delicate process in which the changes in the I-technology-world evolved. I call this type of evolution sedimentation.

SEDIMENTATION

When reviewing the historical variations of the cell phone, it appears that the evolution of the cell phone is not a simple dialectic process.

The characteristics that are added to novel variations do not negate previous ones but rather "sublate" them. Texting did not eliminate voice communications, and voice is still relevant in the age of the iPhone. Neither can I describe the cell phone's evolution in problem-solution terms, as media scholar Paul Levinson does. He names the cell phone "a remedial medium" because, he claims, it solves the problem of being bound to wires at home.[7] While this description may be correct for the first historical variation, it is not necessarily suitable for the fourth variation because the addition of the camera and GPS can barely fit into a problem-solving scheme. Moreover, the reversal of means-end in the relations between humans and modern technologies, as Heidegger and Hans Jonas already pointed out,[8] positions the description of "remedial technologies" as anachronistic and irrelevant for modern and surely postmodern DTs, such as the cell phone.

It may be more accurate to refer to a sedimented model that works in layers, where each new layer adds to the previous and, in a delicate process, shifts the focus to a new type of activity.[9] In such a sedimented way, the transformation of the cell phone from a voice-centric mobile telephone into a writing machine and later into a true multistable device has also transformed our meaning of being-in-the-world.

The fact that a variation's characteristics are not eliminated but remain viable for the next historical variation makes it more difficult to clearly differentiate the start and end points of each historical variation. The modern age has left us with machines such as cars and trains, as well as tools such as hammers and needles. What defines the DT age is not that previous technologies have been eliminated, but that new types of technologies that transform our being-in-the-world have been added. Hence a variation can be identified by an assemblage of features, characteristics and common sets of usages. A sedimented model can serve to depict the assemblage and to show how each new layer adds to previous layers. Through a delicate process, sedimentation shifts the focus to a new type of activity. What is important to remember is that previous layers are not eliminated but rather integrated into additional new variations forming new contexts.

NOTES

1. Cf. Mark Hansen, "The Time of Affect, or Bearing Witness to Life," *Critical Inquiry* 30 (Spring 2004): 584–626.

2. Sherry Turkle has a different view of the role of networked technologies. In *Alone Together: Why We Expect More from Technology and Less from Each Other* (Basic Books, 2012) she asserts that they produce the effect of being "utterly alone" (154). She prefers the critical (and even dystopian) approach to contemporary

technologies, and whenever a respondent brings an example of how technologies enable her to connect to others, Turkle attempts to show that in fact the respondent felt lonelier. Moreover, her descriptions of non-technological loneliness are positive and attempt to impute to loneliness an opportunity to think or to interact with strangers. She overlooks the pain of loneliness among members of the modern urban crowd and the difficulty to flourish and develop positive thinking in this state (cf. p. 155).

3. It should be reminded that a touch screen offers just one kind of a tactile experience, that of a sleek surface.

4. *Businessweek editorial,* 1975, "The Office of the Future," Business Week (2387): 48–70, June 30, 1975, http://www.businessweek.com/stories/1975-06-30/the-office-of-the-futurebusinessweek-business-news-stock-market-and-financial-advice (accessed 28 November 2012).

5. The parking tickets can be bought in metropolitan areas via the cell phone. When the inspector checks a car, she has to type in the plate number and a cellular system confirms that the car is linked to the cell phone from which payments for the parking were made.

6. Osnos, Peter, 2012, "In China, 25 Million People Use Only Their Cell Phones to Read Books," *The Atlantic,* November 27, 2012, http://www.theatlantic.com/technology/archive/2012/11/in-china-25-million-people-use-only-their-cell-phones-to-read-books/265622/ (accessed 28 November 2012).

7. Paul Levinson, *Cell phone: The Story of the World's Most Mobile Medium and How it Has Transformed Everything!* (New York and Basingstoke: Palgrave MacMillan, 2004), 9.

8. Martin Heidegger, "The Question Concerning Technology," in *The Question Concerning Technology, and Other Essays*, transl. W. Lovitt (New York: Harper and Row, 1977): 3–35; Jonas, Hans, "Towards a Philosophy of Technology," In *Philosophy of Technology: The Technological Condition*, ed. Robert C. Scharff and Val Dusek (Malden, Oxford, Victoria: Blackwell Publishing, 2003), 191–205.

9. Cf. the Husserlian model of sedimentation as described by James Dodd, *Crisis and Reflection: An Essay on Husserl's Crisis of the European Sciences* (Dordrecht: Kluwer Academic Publishing, 2004).

Part II

INVARIANTS

Chapter 5

First Invariant

Wall-Window-Screen[1]

WALL-WINDOW

Rich Ling opens his book *New Tech, New Ties: How Mobile Communication Is Reshaping Social Cohesion* with a story of a plumber who enters Ling's home to fix a leaky faucet. The plumber enters while talking on his cell phone, and while ignoring Ling starts working without saying anything. For Ling, this anecdote represents the cell phone's ability to separate the plumber's immediate environment (a home with a leaky faucet) and the space of his personal connections (the plumber's friend, colleague or wife). The cell phone enables the plumber to make such a separation and to erect an imaginary yet effective "wall" between him and his surroundings. With the cell phone, his attention is split between the here-and-now space of Ling's home and a not-here space of personal connections. This space of not-here/not-now is accessible because the cell phone acts as a "window." Although a wall-window may sound at first like an oxymoron, modern architecture supplies us with real examples of wall-windows. Many buildings nowadays feature screen-walls—transparent floor-to-ceiling walls that prevent heat and noise from entering, but let in daylight and views from the outside.[2]

The wall-window model exposes how the cell phone simultaneously separates the user from the physical surroundings (the wall) *and* connects him or her to a remote space (via the window) where the interlocutor resides. Using earphones assists in blocking the auditory inputs from the surroundings and contributes to the construction of this wall. Sometimes, the window manifests its visual aspects when we focus on the cell phone's screen as a visual framing, displaying selected parts of the remote space.[3] Unlike computers and televisions, the cell phone can provide the interlocutor (who resides in the other space) with a window to the user's current physical space through the cell phone's built-in camera, an activity known as video conferencing. Such activation of the built-in camera turns the cell phone into a two-way window, similar to a window in a building that enables people outside the building to look inside into a specific room.

The wall-window model is not reserved for eccentric situations like the one described by Ling in which the plumber is so busy with his cell phone that he starts working without even saying hello. The model is relevant for most people in everyday situations, such as commuting on a train, where the cell phone keeps us within a virtual (private) territory while helping us avoid feelings of loneliness or boredom. Being "transported" to a remote space is sometimes considered hazardous, as in the case of driving while talking on the cell phone.[4]

The wall-window model also reveals how the cell phone can be used to undermine power relations, as in the case of students in class or workers at call centers sending text messages. Harvey May and Greg Hearn argue that as students and workers manage to shift their attention to another space, they are no longer fully subjected to their professor or boss.[5] The cell phone as wall-window empowers them because it enables them to remain physically in the space in which they are fully subjected and yet part of them is elsewhere in the virtual space in which the text messages are exchanged. This example demonstrates how the remote space of the interlocutor is not necessarily the space where he or she resides but can be an abstract and virtual space in which the conversation "takes place."

In this chapter, I show how the wall-window is an invariant of the historical variations of the cell phone. This invariant can be applied for all screens; it is, I argue, not only an invariant of the cell phone but also of media and new media (i.e., media where computers are involved). I further maintain that the cell phone and new media conflate and create a "new new" media, where users are involved in the production and selection of media. The technologies of cell phones, screens and (traditional as well as new) media all serve to disassociate to various extent humans from their physical environments, while linking them as users to another space—be it social, distant or virtual.

The Thickness of the Wall, the Transparency of the Window

It should be noted that not every use of the cell phone results in a thick isolating wall. Sadie Plant characterizes three possible responses to a phone call received in public.[6] The first is "flight," in which the user escapes from the surroundings and looks for a quiet spot. The second is "suspension," in which the user physically remains where she or he is, but stops paying attention to the surroundings and suspends his or her activity conducted prior to the call. The last is "persistence," in which the user remains an active participant in and with the surroundings while at the same time maintaining the phone conversation (voice or text). The first two responses create walls of various "thicknesses"—that is, isolation from the public. Such isolation is not absolute. Ling's research suggests that even in these cases the user is still aware of what is going on around him or her.[7] The third response, "persistence," is a true effort of the cell phone user to remain part of and participant in the environment. It is an effort because each space requires attention. If the two spaces collide so that the attention to one is at the expense of the other, then the result is distraction.

Ingrid Richardson explores the Latin origins of the word distraction, *distrahere*, and reveals it means pulling in different directions.[8] This etymological analysis confers a positive context to distraction, in which attention is not taken or gone but can also be divided between two spaces. Yet, she notes, we distinguish between the spaces so they do not get mixed up. Such an interpretation of distraction presupposes a one-dimensional concept of attention: as more attention is given to the remote space, less is given to the here-and-now, and vice versa.[9] I prefer the term attention over distraction, because distraction presupposes a given level of attention that can be divided, whereas I conceive attention as flexible, liquid and dynamic.

Instead of referring to one-dimensional attention, in which any attention given in one space is at the expense of the other, I propose a two-dimensional model of wall-window (see Figure 5.1). The wall axis represents the attention to the immediate physical environment, and the window axis represents the attention to the remote or virtual space. Attention can be minimal for both spaces and given possibly to a third space, or maximal to both spaces thereby increased as a whole.

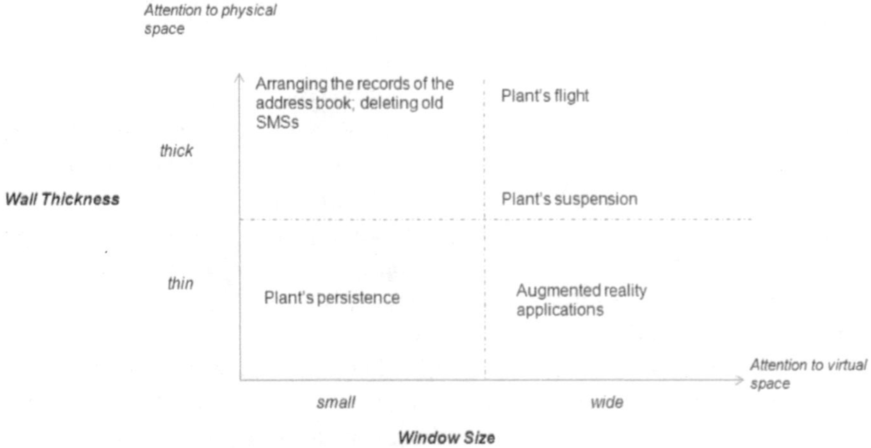

Figure 5.1 The Wall-Window Dimensions. *Source*: Author.

This dual axis diagram accommodates attention/distraction as the size of the window and the thickness of the wall. The smaller the window, the less attention is given to the virtual space; the thinner the wall, the more attention remains in the physical space. As a result, the wall-window model can depict situations where attention to the immediate is distracted, but not to another space, as in the case of a user arranging the records of the address book or deleting old text messages. These maintenance tasks of the cell phone metaphorically build a wall between the user and the here-and-now, while the window to another space is closed. The model can also accommodate augmented reality as described in the fourth historical variation that enhances the here-and-now with more information. Augmented reality can be represented on the wall-window axes as a thin transparent wall because the contact with the here-and-now remains tight and active. It can also be represented as a window of varying size depending on the amount and type of information presented.

The two-dimensional model can also illuminate Plant's third option of persistence where the user continues to perform the here-and-now activities while talking on the cell phone. This option can be modeled as a thin wall with a relatively small window, because the user's attention is equally split between the two spaces. The wall-window model can accommodate situations where people provide attention to two spaces simultaneously, ranging from pilots splitting their attention between the visual stimuli (of the outside world and the instrument panel) and auditory inputs (e.g., air traffic control communications) to teenagers doing homework while the TV is on.[10]

We should remain aware of the possibility that many more axes exist. As Gilles Deleuze and Felix Guattari argue, there can be a thousand plateaus, or axes, to model and describe reality. In this chapter, I concentrate only on two axes: the wall and the window.

The wall and the window axes are not given but produced by the user and the technology. While the window is "provided" by the cell phone technology, the wall is "constructed" by the user. It is achieved through "a face-and-body that says 'I'm on the phone'."[11] This body language places a "wall" between the user and the world, and this wall can be fortified with the assistance of objects like the handset or earphones. Such behavior gave the cell phone its reputation as an "antisocial" instrument.

The wall may be perceived differently by the people around the user, who may feel that public space is being "colonized "or "invaded" by private communications.[12] While the user feels "at home" in the public space when making a private call "behind the wall," the other inhabitants of that space do not share this feeling. On the contrary, for them the wall may not exist at all. Sometimes it is appropriate for people around the user to exercise "civil inattention."[13] thereby contributing "a stone in the wall." Yet, from a social point of view, the participation in the construction of the wall on the part of the people around the user has more than one meaning because body language is not unequivocal. When a user looks for a quiet spot to make a phone call in public, is it considered a display of "courtesy, etiquette and manners,"[14] as part of "strategies to minimize disruption to public spaces,"[15] or is it interpreted as a selfish antisocial act intended to gain privacy? My point here is that the wall can be viewed in various forms by the user constructing the wall and by the others outside the wall.

The body language of being-on-the-phone is not limited to voice communication. Richardson refers to a variety of portable devices, including game consoles, walkmans and iPods.[16] She notes that such devices are characterized by "face work," which signals to the surroundings a sense of being-busy. Likewise, the cell phone can be used to avoid feelings of loneliness and boredom, not only when used for voice communication, as represented by the first historical variation, but also when used for gaming and listening to music (second and third variations), or geo-tagging (adding information as described in the fourth variation thereby turning the world into a more meaningful environment). This extension from voice to gaming, music and augmented reality means that the window invariant does not require a person on the other end; the window can be open to a virtual space of applications, such as games, videos and eBooks.

SCREEN-WALL-WINDOW

Unlike landline telephones, cell phones had a screen from a relatively early stage. As cell phones developed, the screen became a central element. In this section, I examine how the addition of a screen to the cell phone affects the wall-window model. What is a screen? According to Lucas Introna and Fernando Ilharco, the origins of the word can be traced back to the Middle Ages when it meant a shield or a barrier of some kind.[17] Today the word screen has taken on an additional meaning of filtering and even distortion.[18] All of these separation qualities are shared by the wall, since both the cell phone-as-a-wall and the screen separate the user from the surroundings. And yet, the screen shows information in a frame, thereby sharing qualities with a window. The cell phone's screen contains the two aspects of wall and window. Moreover, these aspects are even further accentuated than a standalone screen because the cell phone itself functions as a wall-window.

This can explain why the screen has become an important part of the cell phone. If you think of a cell phone as a telephone with mobility, it is not very clear why a screen is required. The auditory capabilities do not require visual aids. It is a "nice to have," but not a "must." Yet, the cell phone's screen evolved throughout the four historical variations. Back in the first variation, the simple, single-line LED screen already displayed the phone numbers of incoming calls and some minimal form of text. In the second variation, the screen evolved into an alphanumeric display that showed several lines of text in black and white. The third variation was characterized by a colorful screen as large as the palm of the hand and presented multimedia content. The fourth variation requires a high-quality screen on which the augmented reality can be displayed.

In parallel with the growing importance of the screen within the cell phone technology, screen size has constantly increased. This increase occurred in line with the growth of contemporary televisions and flat-screen displays,[19] and the increasing role of the screen, as reflected in Lev Manovich's term "the society of the screen."[20] Table 5.1 shows how the cell phone's screen resolution and size have grown for Nokia's handsets over 14 years. The table depicts cell phone screen sizes (in pixels) by the year of each handset's release. It shows how the screen size and resolution grew from 84 by 48 pixels in 1997 to 640 by 360 pixels and more in 2010. The one exception is the Nokia 9000 series (also known as the "Communicator") that included a relatively large screen of 640 by 200 pixels as early as 1997. By comparison, iPhone's screen size is 320 by 480 pixels (for the first iPhones as of 2007).[21] The Communicator can be regarded as the harbinger of the arrival of larger screens to cell phones and for applications. In 1998, a year after the Communicator's launch, the industry analyst firm Gartner forecasted that "users will demand larger screens to display several lines of data. Current terminals . . . will no longer be acceptable except for users that want a voice-only phone."[22]

Today's screens display not only several lines of data, but also maps, pictures and video. Screens continue to grow in size, culminating in Apple's iPad, which has been dubbed a "fat iPhone without the phone."[23] The iPad looks like an iPhone with a much larger screen that extends beyond the palm of the hand and is enlarged to 1024 by 768 pixels (or from 3.5 inches to 9.7). With its 3G support, it blurs the lines between tablets, personal computers and cell phones. Yet, the iPad is not a cell phone, and is not intended to replace it. In a way it marks the limits of the cell phone's screen, and encourages us to think of the horizon of possibilities for using cell phones as portable computers.

At this point it becomes evident why the screen of the cell phone has evolved from a single LED line to the size of the palm of the hand. As the screen grows in size and improves in quality (more colors, better resolution), its traits as wall and window are both radicalized. The potential isolation from the surrounding environment increases, with more sophisticated handsets and earphones, while the plethora of multimedia applications is opening a wider and richer window. Yet, most scholars have referred to the screen as either a wall or a window.

Table 5.1 Nokia's Popular Screen Resolution and Size (in Pixels), 1997–2010 (2010: Partial Data)

Screen size	*84 × 48*	*96 × 65*	*96 × 68*	*128 × 128*	*128 × 160*	*208 × 208*	*240 × 320*	*320 × 240*	*352 × 416*	*360 × 640*	*640 × 200*	*640 × 360*
1997	4										1	
1998	9										1	
1999	7	1										
2000	5	4									1	
2001	11	3									1	
2002	14	13		3							2	
2003	2	19		14	2							
2004		9		15	6		1				1	
2005		10		9	10	3		3	1		1	
2006			4	2	11	1	6	16	3		1	
2007			3	1	8		10	12				
2008			1		6		17	6		1		
2009			1		4		16	7		3		1
2010			1		4		6	2		2		5

Source: Table constructed by the author based on information from About.com and Manuals Online.

Screen-Window

The window is a popular metaphor for the screen,[24] because both window and screen allow us to see the world through a rectangular frame. The limits of the metaphor become clear when we consider the direction of information flow. Most architectural windows allow flows in and out. However, the cell phone screen, in most cases, is a one-way membrane that allows information to flow to the user. The exception is the use of the cell phone camera, where the user's image and background surroundings are transmitted to the interlocutor's space.[25]

Anne Friedberg seeks an alternative to the window metaphor for the screen, and suggests construing the screen as a virtual window.[26] In her framework, virtuality is a complementary aspect of the real and is not in opposition to it. The virtual is a representation of different materiality and reality, "a second-order materiality."[27] Put differently, it is a reality located in a different space.

This understanding of virtuality enables Friedberg to name the screen a virtual window which can function as a substitute for the architectural window. It can be a substitute for a "real" window because the screen can open for us another space not just metaphorically. When the screen as a virtual window opens another space it complements our here-and-now reality. If Friedberg's framework is applied, then the cell phone can be constructed as a virtual window; it opens another space that is not material but quasi-real. Although she emphasizes the visual aspect of the screen, the cell phone also includes auditory elements. Thus, the cell phone expands Friedberg's virtual window by interweaving the auditory with the visual.

Screen-Wall

Friedberg maintains that the screen as a virtual window "is reliant not on its transparency but on its opacity."[28] The virtual window for Friedberg also expresses opacity, which I describe as the "wall" qualities of the screen. However, as a wall, the screen is more than a metaphor because it has an effect on the world. If as a window it was construed as virtuality, in its capacity as a wall it has actual outcomes. This is, for example, Virilio's approach: "In a certain sense, the screen became the last wall. No wall out of stone, but of screens showing images. The actual boundary is the screen."[29] Virilio emphasizes the role of the screen in contemporary everydayness and its function as a wall. John Armitage explains that the screen for Virilio is the "last wall" because "the geographical difference between 'here' and 'there' is obliterated."[30] While the total obliteration of the gap between the here-and-now and the remote space may be in dispute, there is a consensus that this gap is narrowing and that the cell phone and its screen are contributing to this reduction.

I argue that the cell phone's practices construct an invisible wall of various levels of thickness that denotes a range of separation intensities. By the same token, the screen becomes a wall through the social context of its use.

A phenomenological observation of the screen reveals, according to Introna and Ilharco, the ways in which a screen socially becomes a wall against the environment.[31] The wall comes out as soon as we push the "On" button. Introna and Ilharco describe TV and computer screens that require the users to "sit down, quit—physically or cognitively—other activities we may have been performing, and watch the screen."[32] This, I think, does not apply any longer to contemporary TV and computer experience. Many of us have these screens operating in the background, and not always in the foreground. The cell phone's screen can be regarded as a step back because it requires our attention. It manages to do so because the cell phone's screen "is the place, location, setting, scene, intentional experience of consciousness in which what is relevant or supposedly relevant for us at that particular time is happening."[33] The screen, including the cell phone's screen, not only transforms the way we perceive the world but also which world is to be perceived; it directs our attention, view and perception when it functions as a wall.

The analogy between a cell phone's screen and a wall is contested by Richardson.[34] She contends that there is no separation of the user from the here-and-now because the cell phone "is always-already surrounded by other objects and activities within the spatial topography of the built environment." In a single axis landscape the "being surrounded" blocks the possibility of a split between the surrounding and the remote space. The double axis model enables the coexistence of Richardson's conceptualization of "being surrounded" and the ability of the cell phone (and its screen) to transfer parts of our attention to another space. Our engagement with the surroundings is not a constant but more of a vector of "thickness" of the wall. When it is "thin," we experience the environment as inseparable as described by Richardson; when it is "thick," we tend to ignore the surrounding objects and activities.

Richardson further contends that the cell phone does not require long spans of attention, because "our 'turning towards' [the cell phone's screen] is usually momentary (checking for a text or missed call) or at most can be measured in minutes."[35] This is true for a user immersed in the surroundings. But the wall-window is not a concrete wall; it is invisible and has a social context that gives it flexibility. That is why the wall can last any span of time: a texting student can "return" to her class within a few seconds, while a voice conversation or a game can take her attention for more than an hour.

MOBILE MEDIA

Is the split of attention unique to cell phones? Are there other technologies that provide this separation of attention between two spaces? Apparently, all media technologies—from the printed book to television and video games—dislocate and split the user's attention. When we are immersed in the plot of a novel most of us are less attentive to the happenings around us; when my children watch television, they don't hear me, let alone listen to me; and when people play on the Internet or with game consoles (like the Xbox, Wii or PlayStation) they tend to ignore any interference.[36] An extreme example may be the Walkman, which does not enable its user to hear the surroundings, thereby producing a vocal wall that blocks all auditory inputs from the environment.

I developed the wall-window invariant to represent the ways in which the cell phone modifies our everyday experience. A second look suggests that this invariant may apply to additional technologies, such as media technologies. Another look reveals that this model may apply even to language technics. For example, a teacher in a classroom teaching history provides a wall-window for the students; a conversation between friends in a coffee shop produces a wall-window; even the carpenter using a hammer may be caught in a daydream. These are different forms of wall-windows. The dislocation of attention can happen when our imagination works, or when we talk with other people. While imagination and fantasies come and go, sometimes unintentionally, media technologies ensure this "magic" always happens, whether we are alone or in the company of others. All we need to do is operate these technologies correctly, that is, open a book, turn on the television set, or enter a website.

Many scholars have referred to the cell phone as media.[37] As media, it is a tool for mass communications in a new form. The resulting mobile media includes a variety of formats: SMS, news, voting in reality shows on TV, and mobile music. The list remains open. Such a broad definition of mobile media transforms the cell phone from a mere communication device to a multimedia platform enabling a wide variety of usages and functions. This transformation follows another transformation that has been observed by Roger Silverstone and Knut H. Sorensen.[38] They argue that the addition of the letter "e" as a prefix to words represents the move from information to communication. I would like to suggest that the pendulum may be moving back to information with the prefix "i" (as in the case of iPhone), which has an additional meaning of the first person, usually associated with self-centered activities such as games, navigation, and music.

The prefix "e" is tied to new media; the suggested shift to the prefix "i" may suggest an even newer new media. In other words, the cell phone is not just "media" of mass communications (i.e., printing press, radio and television); it is even "new media," of digitized communications and contents; it is new "new media" that brings new qualities to both traditional and new media by enabling communications everywhere and in novel forms and formats. New new media's ubiquity results in new contents that originate from new sources ranging from short texts providing live updates on events to pictures and videos taken via the cell phone's built-in camera. In addition, the contents (old and new alike) are consumed differently and shift structures and categories. This is evidenced by television programs that urge the viewers to text-in their votes, or by new forms of music consumption, to name only two examples.[39]

New media gained its new-ness because of the deep involvement of computers. Manovich claims, "the logic of a computer can be expected to significantly influence the traditional cultural logic of media; that is, we may expect that the computer layer will affect the cultural layer."[40] From this perspective, a new trajectory may be revealed (see Figure 5.2) stretching from reality (of traditional media), through virtual reality (enabled by new media's computerization), to augmented reality (which is the result of mobile media, as I claim in the fourth historical variation). It is a reality that interacts with the user, and where the user's action may have performative results in the real world.

The common denominator of media technologies—whether traditional, new or "new new"—is their capacity to dislocate our attention to other places (remote in place, time or imaginary) while our physical body remains in full engagement with the surroundings. Media technologies correspond to the wall-window invariant.

The cell phone "redirects" some senses, thereby mediating a different worldview for us. This is what Ihde calls "quasi-illusion," or "subduing the other sensory dimensions."[41] I suggest that this quasi-illusion is not limited to state-of-the-art virtual reality or complex scientific machinery, as originally intended by Ihde, but can be extended to everyday situations where media is involved. The cell phone and media technologies ensure the emergence of the wall and the window. With the screen embedded into the cell phone, the displacement can be not only auditory but also visual, impacting our very embodiment.

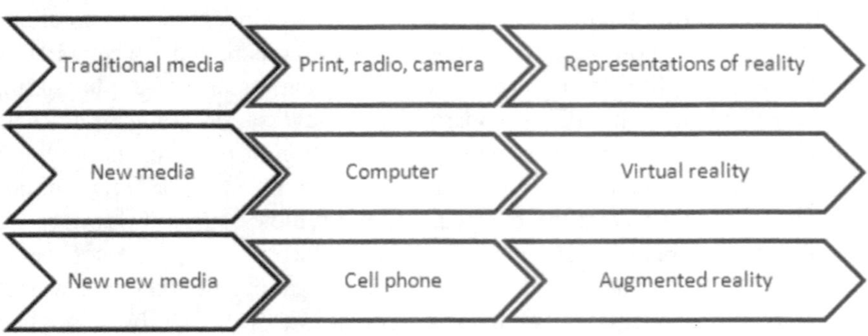

Figure 5.2 Media, New Media, "New New" Media. *Source*: Author.

EMBODIMENT

Inspired by Merleau-Ponty's phenomenology and his multidimensional notion of the body, postphenomenology views the human usage of technology not as a relation between a detached Cartesian ego and an object, but rather as the integration of a technological artifact and the lived body. In this section I propose to think of the split of wall-window in terms of split of embodiment.

Postphenomenology models several types of relations between people, technologies and the world. One of them is the embodiment relation which stands for the incorporation of technologies into our perception in a sense that "we experience [them] as taken into our very bodily experience."[42] A common example would be the eyeglasses that transform the wearer's experience of the surrounding world, while being perceived as part of the wearer's body. Postphenomenological embodiment has two constituents: a Body One that "is the existential body of living, here-located bodily experience, the sense of body elicited by Husserl as Leib, but much better descriptively developed by Merleau-Ponty as the 'corps vécu';"[43] and a Body Two that is "the cultural or socially constructed body . . . the body of the condemned in Foucault, the body upon which is written or signified the various possible meanings of politics, culture, the socius."[44] While Body One represents the experiencing body that inhabits the present, Body Two—requiring a Body One as a pre-condition—is inspired by Foucault and represents the cultural body. Body Two complements Body One by referring to the cultural, social and political aspects. Perceiving a hammer only as a tool for nailing relies on Body One while neglecting Body Two. Such a one-dimensional approach may blind us to seeing other cultural usages for a hammer, such as a paper weight or a weapon, depending on the cultural and sociological contexts. The structure of Body One and Body Two conceptualizes our relations with technologies through the duality of the living-sensing experience and the cultural experience.

The duality of Body One and Body Two shares some similarities with the two spaces of attention modeled by the wall-window invariant: an attention to the here-and-now and an attention to a constructed space. First is the conceptual split that produces a sum larger than its parts. People talking on the cell phone can still participate in the conversation going on in their physical vicinity, as demonstrated by Plant's notion of persistence, in which the user remains attentive to both spaces. It is not an "either-or" relation, but an "and-and" that covers the physical and the remote spaces. Likewise, the theoretical split between Body One and Body Two assists in the conceptualization and understanding of "us" and how we are positioned in the world. Second is the distinction between the here-and-now of the bodily experience and the other space, be it social (in the case of talking on the cell phone with another person) or cultural (as in the case of browsing the Internet through the handset). This is the distinction that Merleau-Ponty and Foucault (and many of their followers) have tried to conceptualize. Notions like interpellation demonstrate the split between one's own needs and the socially-culturally-imposed drivers of human action. Third, the invisible wall, with the help of tangible artifacts such as the handset and earphones, is tightly related to the "natural" or "bodily" aspect of Body One. It is reflected in Virilio's rejection of the "metaphorization" of the screen-as-wall. This aspect of the wall is complemented by a window that is metaphorical or virtual, and tends toward the cultural-social dimensions of Body Two. In line with Friedberg's notion of virtuality as "a second order of reality," the other space is not physically "attendable," but exists in a cultural-social dimension. It is the constructed Body Two that is presented to the other that inhabits the virtual space.

Postphenomenologically, the split of attention between the here-and-now and the remote space may be conceptualized as a split in our embodiment, so that part of the perceptions and experiences remain attached to the space where my physical body is, and another part is "transferred" to the remote space. Yet, it is difficult to draw a clear line between the bodily perceptions that remain in the physical space and those that are transported (by the technology) to the remote space. Likewise the cultural-social aspects cannot be cut into two. This difficulty is partly due to the fact that the two constituents (of attention as well as of embodiment) are interrelated and non-reductive. Combined, they represent *additional* experiences and perceptions that could not have existed without technologies such as the cell phone.

The notions of Body One and Body Two help us configure a split of attention that is richer than a simple divide between visual and auditory stimulus. It can explain how we can read a text message while walking down the street or talk on the cell phone while driving: Body One has a sensory understanding of our immediate surroundings, while Body Two can read a text message, hear music, or connect to a remote space.

THE CELL PHONE MODEL

The model sketched above has four components: wall, window, screen, and media (see Figure 5.3). The relations between the components illuminate certain aspects of the cell phone. Wall and window are attributes that are linked as complementary relations to the here-and-now and remote spaces; the technological components of screen and media are so widely discussed[45] that their relations have become trivial. The screen has been "blamed" for separating the viewers from their immediate environment.[46] Media is often perceived as a window.[47] The cross-links between window and screen or media and wall

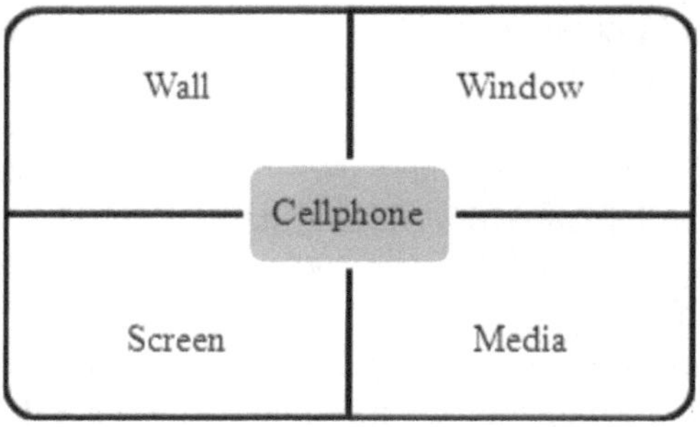

Figure 5.3 Cell Phone's Model. *Source*: Author.

were also mentioned here. And yet the emergence of the cell phone in the middle is surprising. We tend to think of our cell phone as a technology or a social phenomenon,[48] but rarely think of it as media, a screen, an invisible wall or a virtual window. These four aspects, termed here "the fourfold," give us a new perspective for embodiment and attention in terms of wall-window. It helps us understand why the screen of the cell phone grew from a single LED line to the size of the palm of the hand; why some people do not refer to their cell phone as a medium for talking, but more as a tool to consume music or other media; and the feeling of confidence we have when walking in a dark street (wall) with the cell phone open (window), making friendly noises (media) and illuminating our way (screen).

NOTES

1. A version of this chapter was published as an article. Wellner, Galit, "Wall-Window-Screen: How the Cell Phone Mediates a Worldview for Us," *Humanities and Technology Review* (2011) 30: 87–103.

2. See Anne Friedberg, *The Virtual Window: From Alberti to Microsoft* (Cambridge and London: The MIT Press, 2006), 16–17. Recently, such wall-windows are used also to display images to passerbys. When the image is viewed from the inside, such a window is marketed as "transparent OLED" or "transparent LCD," depending on the display technology.

3. See Lucas D. Introna and Fernando M. Ilharco, "Ontological Screening of Contemporary Life," *European Journal of Information Systems* 13 (2004): 221–234.

4. See the debate over this question in the collection of articles in a special issue of *Techné* Vol. 18(1/2) (2014).

5. Harvey May and Greg Hearn, "The Mobile Phone as Media," *International Journal of Cultural Studies* 8, no. 2 (2005): 195–211.

6. Sadie Plant, *On the Mobile: The Effects of Mobile Telephones on Social and Individual Life* (Motorola, 2003), http://www.motorola.com/mot/documents/0,1028,333,00.pdf.

7. Rich Ling, *The Mobile Connection: The Cell Phone's Impact on Society* (Morgan Kaufman, 2004), 135–137. Others like Sherry Turkle see only the isolation: *Alone Together: Why We Expect More from Technology and Less from Each Other* (Basic Books, 2012), 155, 161.

8. Ingrid Richardson, "Mobile Technosoma: Some Phenomenological Reflections on Itinerant Media Devices," *Fibreculture*, 2005: http://journal.fibreculture.org/issue6/.

9. According to this logic attention cannot be divided, see Robert Rosenberger, "Embodied Technology and the Dangers of Using the Phone While Driving," *Phenomenology and the Cognitive Sciences* 11, no. 1 (March 2012): 79–94.

10. Cf. N. Katherine Hayles, "Hyper and Deep Attention: The Generational Divide in Cognitive Modes," *Profession*, 2007: 187–199; Bernard Stiegler, *Taking Care of Youth and Generations* (Stanford: Stanford University Press, 2010); Galit Wellner, "Multi-Attention and the Horcrux Logic: Justifications for Talking on the Cell Phone While Driving: Justifications for Talking on the Cell Phone While Driving," *Techné: Research in Philosophy and Technology* 18, no. 1–2 (Winter-Spring 2014): 48–73; Marcel O'Gorman, "Taking Care of Digital Dementia," *CTheory* (http://www.ctheory.net/articles.aspx?id=740, accessed 24 May 2015).

11. Ingrid Richardson, "Faces, Interfaces, Screens: Relational Ontologies of Framing, Attention and Distraction," *Transformation* 18 (The Face and Technology) (2010), http://www.transformationsjournal.org/journal/issue_18/article_05.shtml.

12. Hans Geser, *Towards a Sociological Theory of the Mobile Phone*. 3.0. http://socio.ch/mobile/t_geser1.pdf. 2004.

13. Leslie Haddon, "Cultural Differences in Communication: Examining Patterns of Daily Life," *Mobile Communication and Social Change* (Seoul, Korea, 2004).

14. Rich Ling, "The Social Juxtaposition of Mobile Telephone Conversations and Public Spaces," *The Social Consequences of Mobile Telephones* (Chunchon, Korea, 2002).

15. Haddon, "Cultural Differences."

16. Richardson, "Faces, Interfaces, Screens."

17. Introna and Ilharco, "Ontological Screening."

18. Friedberg, *The Virtual Window*, 17.

19. Ibid., 138.

20. Lev Manovich, *The Language of New Media* (Cambridge and London: The MIT Press, 2001), 94.

21. Although similar in screen size, the Communicator weighed almost 400 grams, while the iPhone weighs less than 140 grams.

22. Phoebe Leet, "The Mobile Phone: What Will Become of It?" *Gartner/Dataquest* MOBD-EU-DP-9810 (1998, October 28), p. 8.

23. Rik Myslewski, "Apple's iPad - Fat iPhone without the Phone: The tablet that can't multitask," *The Register* 27 January 2010 (http://www.theregister.co.uk/2010/01/27/ipad_comment/, accessed 23 May 2015).

24. For example, Manovich, *The Language*, 94–115; Don Ihde, *Bodies in Technology* (Minneapolis, London: University of Minnesota Press, 2002); Friedberg, *The Virtual Window*, 15–18; Richardson, "Faces, Interfaces, Screens."

25. This exception is Paul Virilio's rule, according to which the screen has a potential to become a reverse window through which we can be watched by others. See

Paul Virilio, *The Vision Machine* (Bloomington & Indianapolis: Indiana University Press, 1994).

26. Friedberg, *The Virtual Window*. Furthermore, she warns us not to confuse digital and virtual (10).

27. Friedberg, *The Virtual Window*, 11.

28. Ibid., 138.

29. Virilio, *The Vision Machine*, 181.

30. John Armitage, "Beyond Postmodernism? Paul Virilio's Hypermodern Cultural Theory," *CTheory*, www.ctheory.net/articles.aspx?id=133 (2000).

31. Introna and Ilharco, "Ontological Screening."

32. Ibid., 225.

33. Ibid., 225.

34. Richardson "Mobile Technosoma."

35. Richardson, "Faces, Interfaces, Screens."

36. See Geser, *Towards a Sociological Theory*, 8–9.

37. For example, Maurizio Ferraris, "Where are You? Mobile Ontology," *Seeing, Understanding, Learning in the Mobile Age* (Budapest, 2005), http://www.fil.hu/mobil/2005/Ferraris.pdf (accessed 22 October 2010); May and Hearn, "Media"; Gerard Goggin, *Cell Phone Culture: Mobile Technology in Everyday Life* (London and New York: Routledge, 2006); Gerard Goggin, "Adapting the Mobile Phone: The iPhone and its Consumption," *Continuum: Journal of Media & Cultural Studies* 23, no. 2 (2009): 231–244.

38. Roger Silverstone and Knut H. Sorensen, "Towards the 'Communication Society'," in *Media, Technology and Everyday Life in Europe: From Information to Communication* (Hants and Burlington: Ashgate, 2005).

39. Mike Featherstone, "Ubiquitous Media: An Introduction," *Theory, Culture & Society* 26, no. 2–3 (2009): 1–22; Goggin, "Adapting."

40. Manovich, *The Language*, 46.

41. Ihde, *Bodies in Technology*, 38.

42. Don Ihde, *Postphenomenology and Technoscience: The Peking University Lectures* (New York: State University of New York Press, 2009), 42.

43. Ihde, *Bodies in Technology*, 69.

44. Ibid., 70. This distinction may remind the reader of the relations between flesh and body in Merleau-Ponty's or between sex and gender in Judith Butler's *Gender Trouble: Feminism and the Subversion of Identity* (New York and London: Routledge, 1990). Similarly Ihde makes an analogous distinction using other terms, such as micro-perception and macro-perception, see Don Ihde, *Technology and the Lifeworld: From Garden to Earth* (Bloomington and Indianapolis: Indiana University Press, 1990). These components of embodiment can also be found in Rosi Braidotti, *Nomadic Subjects: Embodiment and the Sexual Difference in Contemporary Feminist Theory* (New York, Chicester, West Sussex: Columbia University Press, 1994).

45. For example, Friedberg, *The Virtual Window*; Manovich, *The Language*.

46. For example, Introna and Ilharco, "Ontological Screening."

47. For example, Friedberg, *The Virtual Window*.

48. For example, Ling, *The Mobile Connection*; Rich Ling, *New Tech, New Ties: How Mobile Communication Is Reshaping Social Cohesion* (Cambridge and London: The MIT Press, 2008); James E. Katz, *Magic in the Air: Mobile Communication and the Transformation of Social Life* (New Brunswick and London: Transaction Publishers, 2006); Haddon, "Cultural Differences."

Chapter 6

Second Invariant

The Quasi-Face of the Cell Phone[1]

THE CELL PHONE'S SCREEN

Have you ever wondered why a cell phone has a screen? While a historical explanation may elucidate how the screen evolved for this particular technological artifact, it is less likely to explain why it got a screen in the first place. The average engineer may answer that there is a need to display digits of a phone number. This answer places importance on the reading of data, yet it ignores the possibility of other means of representation, such as auditory expression. The urge to display data is not unique to cell phones. Lately some artifacts have also acquired screens, like my new refrigerator that has a screen displaying the temperatures of the fridge and the freezer. I can think of alternative means of conveying the message, such as a single LED that denotes a properly working mode, or a dashboard-like indication. For some reason, the screen was selected by the manufacturer in order to display this particular information. Displaying information by means of an indicative LED or a dashboard does not have the same effect as displaying information on a screen. From the user's point of view, screens function like magnets for the sight. One cannot refrain from looking at them, and so we find ourselves staring at the screen of televisions, computers, cell phones and even refrigerators. The screen serves as a form of conveying a message, even in its small form and the digit-only display. Beyond displaying data to the user, the refrigerator "expresses" itself even though the range of expressions is limited. It is as limited as that of the first generation of cell phones whose display was a digit-only screen.

For many years the cell phone's ancestor, the landline telephone, had no screen at all. Yet, the cell phone had a screen almost from its very beginning. The first generation of cell phones was characterized by a modest screen.

It was a digit-only display of eight-to-ten characters, aimed at showing the phone number of an incoming call or the numbers pressed down for an outgoing call. It required some creativity to display letters, words and phrases, as shown in the second historical variation. Since then, the cell phone's screen has evolved into a large colorful "window" that displays texts, photos, videos and maps. Such a display, I argue, enables more than a reading of data—it encourages the user to refer to the cell phone as an "other." The cell phone not only displays various media; it "remembers" contacts, it plays games with the user (sometimes it wins!), or it gives directions where to go. Ihde identifies a technology that conducts a certain human behavior as a "quasi-other," which maintains alterity relations with the user. It means that one relates to a technological artifact not as a mute object, but as an object loaded with some human traits, such as retaining memory, conveying relevant information, or providing entertainment. Put differently, quasi-otherness is that quality in objects which positions the human and the object in a human-like relation. Yet, alterity relations can be maintained not only with cell phones but with any technology. Certain religious contexts supply some extreme examples of how artifacts and objects can be treated as "others."

My refrigerator is not a full-fledged other. It is a form of a quasi-other because it has a screen that engages me as a user. My claim is that alterity relations are augmented when a screen is involved. The screen provides a certain focal point which draws one's attention and "promises" to participate in an exchange with the user. Obviously, the screen is not fully equivalent to a human face, and hence does not evoke noble human values. The technological screen-face is indeed not an ethical face. But human exchange includes layers of communications, not all of which are ethical. So does the face. The screen is a subset of a face. It shares a few elements with the human face. It is a quasi-face.

While the screen relates to the face, other technologies relate to different body organs. For example, the wheel has a particular relation to the human leg; their ends are similar (enabling one to get from point A to point B) but their executions are different (in form, material, control, flow, operation, etc.). The wheel can be constructed as quasi-leg—the wheel is not a leg, nor is it an enhancement of a leg. It relates to the leg in a particular way, which makes it a quasi-leg. In a similar fashion, the screen is not a face, yet it relates to the face in a particular way. It is a quasi-face.

Emmanuel Levinas is widely known for his ethical analysis of the alterity of the Other and the face. In this chapter I turn from the ethical to the ontological and discuss Levinas' tool analysis. I show how this analysis calls for a shift from the hand to the face as the main interfacing organ. If an ethical analysis reserves the face to the human, the ontological investigation allows a gradation in which animals and technologies can participate. Next, I turn to postphenomenology's alterity relations and examine how this gradation works in order to rethink the relations between humans and technologies. Finally, I return to the screen and its quasi-facial quality.

"WE LIVE FROM GOOD SOUP"

Martin Heidegger in *Being and Time* gives an account of being-in-the-world using vocabulary such as thrownness, fear and *angst*. The human being (as *Dasein*) is thrown into the world, finding itself—in the best case scenario—working with a hammer in a workshop. While working, Heidegger's human does not enjoy work, food or hospitality. As Levinas says, "Dasein in Heidegger is never hungry."[2] The setting in *Being and Time* does not invoke any feelings toward objects, let alone other human beings. It is a world of solitude.

Levinas, a student of Heidegger, portrays a different human being that enjoys the surrounding world. Levinas' basic assumption is that "we live from 'good soup,' air, light, spectacles, work, ideas, sleep, etc . . . . These are not objects of representations. We live from them."[3] Heidegger's worker in a Levinasian setting could have smelled the hammer hitting the nail, heard the knock and dreamt of a good soup, all at the same time. Levinas' positive spirit and the emphasis on enjoyment are in striking contrast to Heidegger's negativity.[4] Levinas continues, "The things we live from are not tools, nor even implements, in the Heideggerian sense of the term . . . . They are always in a certain measure—and even the hammers, needles, and machines are—object of enjoyment, presenting themselves to 'taste,' already adorned, embellished."[5] According to Levinas, our relations with things and tools are not only relations of utility but also involve a full sense of our well-being. Through Levinas' critique of Heidegger, a new tool analysis is revealed.

The Levinasian "Tool Analysis"

One cannot easily find Levinas' tool analysis and the reader of *Totality and Infinity* has to look for clues and scattered comments to construct it. Such difficulty may be imputed to Levinas' prioritization of the ethical over the metaphysical, and to the metaphysical over the ontological.[6] Levinas departs from Heidegger mainly by shifting from the ontological to the ethical, but also differs from him in his formulation of the traits of his ontology, especially those that concern technologies.

Heidegger's starting point in *Being and Time* is the ontology of beings. One type of being is the thing, which is exemplified by the hammer. Levinas, like Heidegger, analyzes concrete things, which he defines as "solids with contours clearly delimited."[7] Objects, according to Heidegger and Levinas, are therefore lifeless physical things which humans can see or touch.[8]

Both thinkers move from tools to machines, that is—mechanical technologies. Levinas implicitly follows Heidegger's overlooking of electronic artifacts,[9] thereby concerning himself with mechanical technologies that have no screens, at least not as a main interface. Neither Heidegger nor Levinas analyze screen technologies such as television or cinema. But unlike Heidegger, Levinas' concept of the machine is to some extent utopian, since he considers "the wheels of machinery [to be] perfectly adjusted to one another and form an unbroken continuity."[10] He remains committed to his "good soup" prism through which he looks at the world, and ignores the fact that machines do break down from time to time.[11]

The relation between humans and things, according to Levinas, is one of "living *from*." It is dramatically different from Heidegger's readiness-to-hand (*zuhandenheit*) and presence-at-hand (*vorhandenheit*). What is missing in Heidegger's analysis is that our use of technology is not necessarily a mere means-to-an-end, but also an experience and a sensing of one's being.

Like Heidegger, Levinas also uses the hammer as an example, but with a twist. Levinas portrays enjoyment relations in which the human bearer depends on the tool she deploys for her own well-being-in-the-world. Being-with-a-tool, according to Levinas, is working with a tool while sensing one's body and physical application.[12] A tool is more than something-in-order-to; it is something that makes us feel, be it a sufferance or an enjoyment. The enjoyment is so central in Levinas' conceptualization of tools that he goes on and asserts, "To enjoy without utility, in pure loss, gratuitously, without referring to anything else, in pure expenditure—this is the human."[13] The construct of *Homo Faber* has therefore additional aspects, in which there is no predetermined goal, no in-order-to. These aspects make *Homo Faber* more human. In a sense, it is a meeting point between *Homo Faber, Homo Ludens* and *Homo Aestheticus*. We may enjoy a tool, as Levinas notes, without thinking what it is good for, or its presence-at-hand, as Heidegger does.[14] We can also enjoy an "embellished" hammer.[15] Levinas thinks of the human experience regardless of whether the tool functions or not, while Heidegger thinks that once the tool breaks down it is the tool that we think of, and not our experience with it. Hence, "living *from*" and the accompanying enjoyment open a horizon for feelings and sensations that exceed the Heideggerian in-order-to relations.

Levinas' positive approach to technologies is reflected not only in *Totality and Infinity*, but also in his article, "Heidegger, Gagarin and Us," published in 1961, the same year *Totality and Infinity* appeared. In this article he refers to technologies more explicitly, in response to Heidegger's essay "The Question Concerning Technology." While Heidegger criticizes modern technologies such as the hydroelectric dam on the river Rhine, Levinas is optimistic with regard to modern technologies. He refers the technoscientific achievement of Yuri Gagarin flying to space as an example of modern technologies and writes,

> What is admirable about Gagarin's feat is certainly not his magnificent performance at Luna Park which impresses the crowds; it is not the sporting achievement of having gone further than the others and broken the world records for height and speed. What counts more is the probable opening up to new forms of knowledge and *new technological possibilities*, Gagarin's personal courage and virtues, the science that made the feat possible, and everything which that in turn assumes in the way of abnegation and sacrifice.[16] (emphasis added)

Levinas refuses to follow what he considers the common utopian view of Gagarin's achievements. At the same time, he rejects the Heideggerian dystopian-romantic approach to technology according to which opening to the world equals to a return to nature.[17] Moreover, Levinas criticizes the enframing (*Gestell*) approach in "The Question Concerning Technology," in which modern technologies are responsible for setting the world as a "standing reserve" (*Bestand*). Instead, he sees technology as an opening up of a new horizon for human beings. Whereas Heidegger sees a dangerous future imposed by modern technologies, Levinas is optimistic when he looks ahead because of the "new technological possibilities."[18] It is the difference between the frame and the horizon: the former is limiting the view, the latter remains open and curious toward the future.

To conclude, Levinas rejects Heidegger's early tool analysis in *Being and Time* and specifically the in-order-to approach, as well as the enframing approach in "The Question Concerning Technology." The world according to Levinas is a human-centric world and can thus be brighter and more playful. We can enjoy it with or without utilities. In Levinas' world, what is important is that we can enjoy the act of working and using tools, as well as the aesthetics of a tool or seeing the results of the work. The emphasis is on the positive aspects of the human experience.

Tools and Organs

Departing from a human experience point of view, Levinas makes a distinction between tools and equipment *(Zeug)* on the one hand and things-necessary-for-living on the other. The latter includes "furnishings, the home, food, clothing."[19] Such a distinction makes sense, though he does not explain the difference, nor what it implies.[20] But such a classification is dictated by the in-order-to relations between the user and the technology, a Heideggerian criterion that Levinas expressly rejects. If one follows Levinas' line of thinking, then the classification should be made by the experience produced from the interaction between the human and the artifact. One possible classification can be that of the bodily organ required for the operation of a technology. A hammer requires a hand; food requires a face.

Such a classification reveals another aspect of the chasm between Heidegger and Levinas, reflected in the different organs each thinker makes reference to: the hand for Heidegger and the face for Levinas. The organ with which one approaches technology can define that technology. Therefore, I suggest distinguishing between technologies of the hand and technologies of the face. Technologies of the hand include the tools such as a hammer, a needle and a pen, but also DT like a keyboard, mouse and remote control.[21] Using these technologies require the hand, and to a much lesser extent, the face. I can knit without looking at the needle and type without looking at the keyboard. Technologies of the face include traditional media technologies such as a painting, a theater performance or a book, as well as contemporary media technologies, such as the cinema, TV, computers and cell phones. The construction of wall-window as detailed in the previous chapter is performed mainly through the face. Although these media technologies are facial, they are sometimes combined with hand technologies like a television remote control, a computer's keyboard, or a cell phone's handset.[22] In the next sections, I review Levinas' approaches to the hand and to the face. My review of the former is mostly ontological, the latter is mostly ethical. In the final section I assess the ontological aspects of the face.

THE HAND

For Levinas "the hand is by essence groping and emprise."[23] The hand serves to investigate the material world. One can find three distinct ways the hand is used for investigating the world in *Totality and Infinity*: the usage of objects, labor and ownership.

Levinas describes usage of objects as entangled with the hand, so that "the form of all objects calls for the hand and the grasp. By the hand the object is in the end comprehended, touched, taken, borne and *referred* to other objects, clothed with a signification, *by reference* to other objects."[24] The hand serves to grasp objects in the world and to investigate them. The hand, according to Levinas, is the major organ for interacting and interfacing with objects. It is an organ for using objects. In this sense, Levinas echoes Heidegger who stresses that the relations with objects, be it ready-to-hand or present-at-hand, are relations with the hand. For Heidegger, regardless of the state of the tools, it is via the hand that a relation to an object is established. At this point Levinas seems to be in agreement with Heidegger.

The hand is so central in the operation of the body in the world that Levinas considers the possibility of reducing the body to the hand.[25] Both Heidegger and Levinas maintain the human hand has a prime role in exploring the world and in thinking of the world.[26]

The hand does more than use objects; it is an organ of work that transforms the objects and the world. That was already analyzed by Friedrich Engels who said, "[t]he specialization of the hand—this implies the *tool*, and the tool implies specific human activity, the transforming reaction of man on nature, production."[27] Such a role of the hand characterizes the tool and the early machines as described by Marx and Engels, and is less relevant for factory machines and DTs. By positioning the hand as a means to control nature, Levinas reveals his commitment to a modernist point of view as portrayed by Engels according to which humans developed science and technologies in order to control nature as encoded in the world's elements.

In addition to being the organ that uses objects and performs labor, according to Levinas the hand also establishes ownership. He characterizes ownership as a meeting point between labor and dwelling. In its capacity as the organ of work, the hand needs the dwelling. The hand and the home are co-dependent, so that the hand builds the home, and the home enables the hand to store collections and works performed by the hand. The home stores the products of the work performed by the hand. The hand in this third sense has a central role in the creation of ownership. Levinas writes, "Possession is accomplished in taking possession or labor, the destiny of the hand. The hand is the organ of grasping and taking, the first and blind grasping in teeming mass."[28] According to Levinas, the destiny of the hand is taking possession. He ties labor and ownership together as interdependent functions of the hand. Work that does not necessarily require the hand cannot be classified as labor by Levinas' standards. The hand for Levinas is more than a sensing organ.[29] It is the means to dominate and take possession of the surrounding world. From this commanding and dictating position, the hand (and after it—the body) can use objects, work them and own them. Just as work requires a home to store things, so does possession.

The hand, according to Levinas, has its limitations.[30] The hand uses objects, works them and owns them in so far as it uses them as a means-to-an-end. These ontological descriptions of the hand seem to fall in the Heideggerian pattern of in-order-to. There is a nuance in which Levinas departs from Heidegger, in which the hand is the means and the end.[31] However, even when functioning as an end, the hand is insignificant for the ethical analysis in *Totality and Infinity*. Such an analysis is reserved for the face.

THE FACE

Levinas declares, "The notion of the face differs from every represented content."[32] The face is unlike other representations because the "face-iality" of the face does not lie in its visible qualities. The face is more than a mere collection of eyes, nose, cheeks or mouth which can be seen, touched or represented. It surpasses its material layout.[33] It is a "corporeal emblem," to use Bernard Waldenfels terms.[34] The face, according to Levinas, is what moves and demands a response from the subject, and yet what makes the face move and demand a response remains inexplicable. Because the face is characterized by something that is not known or understood, Levinas uses, according to Jacques Derrida, a series of negations. Levinas, Derrida says, is "progressing by negations, and by negation against negation,"[35] in order to explain that which cannot be explained in the face. Thus, the face is "often quite ambiguous."[36]

Yet, the face plays an important role in the presentation of the other. Levinas writes, "The way in which the other presents himself, exceeding the idea of the other in me, we here name face."[37] As an essential element in the presentation of the other, "What we call the face is precisely this exceptional presentation of self by self."[38] The interaction with the other is based on the face, and hence the face demands an ethical response.[39] This chapter will not examine the ethical features in Levinas' writing. Instead it seeks to reveal the ontological aspects of the face.

Face and Language

Levinas asserts, "The face speaks. The manifestation of the face is already discourse."[40] The face speaks in two complementary senses: in the words composing a language[41] and in the visual representation.[42]

One of major functions of the face is the exercise of language.[43] Susan Handleman stresses the dual nature of language in the context of the intersubjective relations, as a relation that both maintains and bridges the separation between me and the other.[44] Language not only connects people but it also separates.

Language exists and operates in the facial relations, yet it is not limited to a specific organ in the face. It can be the mouth, the ear, or even the eye. Moreover, other organs outside the face are possible sites for language as well. For example, "Gestures and acts produced can become, like words, a revelation."[45] Language, according to Levinas, becomes "not only verbal signs but all signs can serve as language."[46] Language is not limited to spoken or written discourse. The body can talk via the clothing one puts on, and the hands can talk via gestures.[47]

The extension of the site of language from the face to the body broadens the notion of discourse from spoken words to gestures. A further extension is suggested in the shift from bodily organs to externalities such as clothing that serve as signs. Can other functions of the face be delegated to artifacts? Is it feasible to "de-sanctify" the face and leave the ethical level at a somewhat lower ontological level? Can we attribute a face to nonhumans?

The Nonhuman Face

While the face is a major ingredient of the human, things and animals do not have a face, according to Levinas.[48] To him, even pagan gods do not have a face.[49] The face is uniquely reserved for human beings. Levinas scholars like Waldenfels criticize the exclusion of animals from the property of face, yet still consider things, technologies included, as being faceless (see Figure 6.1).

I would like to suggest a more nuanced approach to the face that would allow animals and technologies to possess a variety of facial forms. In this approach, animals would have a face—not a human face, but a different kind of face. Likewise, technologies have a form of face that I term here quasi-face. Just like one cannot resist the human face and cannot refrain from looking at it, so one cannot refrain from looking at the screen of televisions, computers and cell phones. These screens demand our attention. The proposed approach is ontological and is not focused on the unique ethical implications of the human face.[50] The ontological proposition stretches into a composite structure that includes not only the human but also the faceless object, the allegedly faceless animals and technologies which can be understood as having a

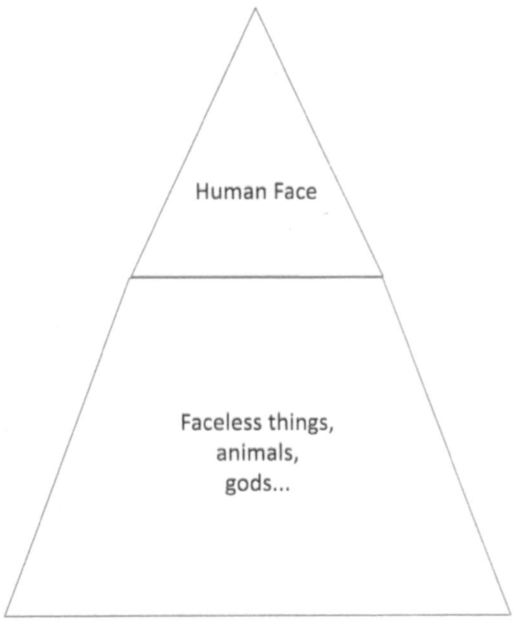

Figure 6.1 The Face Dichotomy. *Source*: Author.

quasi-face (see Figure 6.2). The ethical burden can be allocated along such a structure, in which the human face bares the most significant ethical consequences, and the technological quasi-face has a somewhat reduced ethical call, if at all. The notion of quasi-face softens Levinas' strict and "totalistic" approach to the face that results in a dichotomy of human-nonhuman. Such a structure allows animals and technologies to have permutations of a face by transforming the binary reference to a composite.

A screen of a cell phone is a type of a quasi-face. It opens a window to the face of the other in ways Levinas did not consider. It may allow a discursive exchange with others or provide a reduced version of a face. The screen of the cell phone, like a façade of a home, represents an exteriority which hides an interiority. The screen acts like a face that requires a response, but it is not a face. It is a quasi-face.

POSTPHENOMENOLOGICAL QUASI-OTHERNESS

Ihde seems to agree with Levinas that some technologies are things-necessary-for-living. But instead of getting into classifications, he expands the necessary-for-living to human existence; he maintains that empirically and historically humans cannot live without technologies like language,

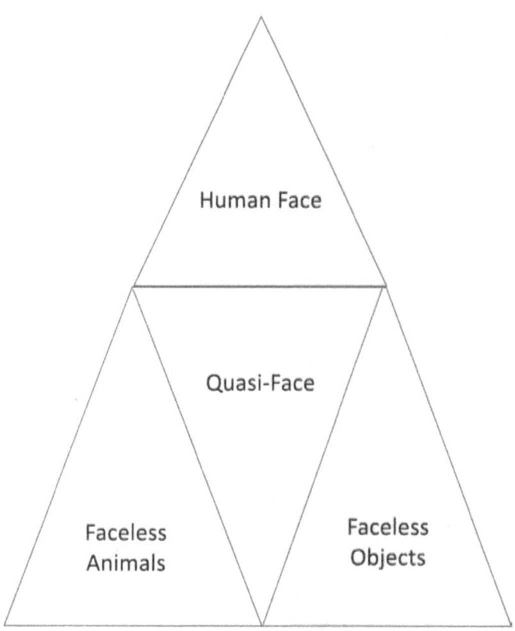

Figure 6.2 "Quasi-Face, Faceless Animals and Faceless Objects." *Source*: Author.

clothing, cooking or habitation. Our experiences of the world are technologically mediated, directly or indirectly.[51]

Ihde identifies four types of relations between humans, their technologies and the surrounding world: embodiment, hermeneutic, alterity and background relations. Often the screen is analyzed through the hermeneutic relations prism, in which it serves to display data that is read by human beings. The approach I suggest here shifts the focus from reading the screen to a more emotional reference to the screen in which the screen and the artifact behind it are referred as an "other." It is a shift from hermeneutic relations to alterity relations in which technology serves as a quasi-other. Ihde explains, "Technological otherness is a *quasi-otherness*, stronger than mere objectness but weaker than the otherness found within the animal kingdom or the human one."[52] The construct of quasi-other is located midway between inanimate objects and living beings. Basically, any inanimate object can be the subject of alterity relations. Children maintain such relations with their toys, whether they have faces, like dolls and teddy bears, or do not have faces, like cars and their favorite blanket. Some religions refer to faceless things as having free will.

Although any artifact can maintain alterity relations with its user, certain technologies provide an opportunity for more intense alterity relations. Consider the case of an interaction with an ATM. One communicates with the technological device as if it were a human teller in a bank. But it is not, it is merely a computerized machine with a screen. It provides communication with a quasi-other. The dialog with the machine produces an augmented form of alterity, whether the dialog is written or spoken (as in the case of telephonic menus).

Ihde constructs a gradation of otherness, topped by the human, followed by animals, then technologies, and finally—objects or things. Within technologies, cinema, television and computers are "dramatic forms" of alterity, according to Ihde.[53] They have "enhanced alterity" in comparison to other technologies because of their ability to display a dynamic picture of the world. These technologies, I argue, have dramatic alterity relations with us, not only because they are dynamic, but also because they have a screen. A computer's and cell phone's screens possess even more dramatic forms of alterity because these technologies are interactive. That is, they maintain a certain dialog with the human operator. This feature allows humans to refer to these technological artifacts in an anthropomorphized way, regardless of the content presented on the screen.[54] Therefore, the cell phone's screen serves as a quasi-face, thereby making our relations with the cell phone more "human" than the relations we maintain with handy tools and other non-screen technologies. Interrelations are constitutive, whether vis-à-vis a human or vis-à-vis technology. Just as the relations with other human beings constitute us in a certain way, so too do technologies co-constitute us in another way.

Screen as Quasi-Face

The concept of quasi-face links a contemporary technological artifact with the face constructing a "meeting point" at the screen. The screen, I argue, acts as a quasi-face. It is not a human face and hence statements like "the screen has become humanized"[55] should be interpreted in a non-anthropomorphized way. Instead of talking in terms of humanization, I prefer the formulation stating that the screen becomes the quasi-face of a technological quasi-other. As quasi-face, the screen and the face have several meeting points:

First, both the face and the screen represent interiority that is not necessarily externalized. Like a foyer of a home which does not reveal the inside, the screen is the exteriority of software codes. If the face is like a theater for displaying feelings and thoughts without exposing the back stage, then the screen is like a theater for displaying multimedia contents without exposing the algorithms operating "behind the screen." In this sense, a "poker face" is equivalent to an empty screen, for it does not show feelings and thoughts, while the interiority remains veiled. In addition, just as one cannot be in touch with the other's interiority, one cannot touch the content that is displayed on the screen. A touch screen, although involving touching, brings to the surface a display of contents while not changing the "back stage."

Second, the screen, like the face, serves as a communicative interface. While the face serves as an *inter*-face point between humans, the screen serves as an interface point between humans and their technologies. It can present the face of another person, distant in place or time; it can display a representation of a collective, as in the case of websites that aggregate contents from many users; and it can represent a technological other, as in the case of a cell phone that greets with the word "welcome" on its opening screen or enables interaction with an imaginary figure like Talking Tom Cat.[56]

Third, both types of interface function on the basis of language. Language enables a dialog, an exchange of thoughts, needs and wishes with the other or the quasi-other. The interaction with the screen amounts to a language in the sense that the screen provides commands and performs certain actions. Obviously there are differences; the most important one is the signification. While human language is expressive in that "discourse founds signification" (as the headline of *Totality and Infinity*'s chapter III.B.4. states), the human-to-screen dialog is designed to be operational. Human language signifies; the screen's language, because of its instructional makeup, is intended to operate applications in an unambiguous, definite way. A quasi-face therefore has a quasi-language. The quasi-language is operational and does not carry any meaning; it is not a condition for thought,[57] nor does it amount to justice.[58] Another difference is the type of reaction produced by the face and the quasi-face. The face responds; the screen is a display of the response, though the displayed responses are not necessarily those intended (see the experience of using a search engine with the expectation of finding a relevant answer, which is not always fulfilled).

Although the face and the screen respond in different modes, neither of them are ever neutral. This is the fourth shared characteristic. The screen-as-quasi-face may explain why we expect the screen to present truth. If it fails to do so, we become disappointed, as though we have confronted a "lying face." Behind the face there is a single subjectivity that operates, while behind the screen there are many humans and nonhumans.

Fifth, the face and the screen extend beyond "physical boundaries." The face's extensions are other expressive organs such as the hands. The screen's extensions are peripherals like the keypad of the cell phone or the microphone. These peripherals enable the interaction with the technological artifact, for example, by typing on a keypad or giving oral instructions via the microphone. An anthropomorphic approach would have to consider the keypad and mouse as hands that write, the speaker as a mouth that makes sounds, the microphone as an ear that can listen to oral inputs and the embedded camera as eyes. These organs assist the artifact in communicating with the user, and in the case of the human, to communicate with the other in a richer way.

Though the screen is comparable to the face, there is one major difference between them. The human face demands an ethical response. Even if the screen may produce such an effect, it is not necessarily designed to be ethical.[59] The ethical aspect of the face exceeds the ontological one.[60] The screen, however, remains on the "ontic" level.

To sum up, both the face and the screen are regarded as displays of an interiority to which we have no access. Neither are standalone entities. While the face is inherently linked to the human other, the screen has recently been added to everyday artifacts. The coupling of a screen to technologies like the cell phone and the refrigerator facilitates the reference to the technological artifact as a quasi-other. It may elevate the relations from the human-thing level to the quasi-inter-subjective level turning them into alterity relations.

THE CELL PHONE'S QUASI-FACE

The question of why the cell phone got its screen is the main concern of this chapter. A hermeneutic agenda—that is, the presentation of data to be read by the user, could have been the original motivation for the inclusion of the screen in cell phones. However, it may turn out that artifacts enhanced by a screen tend to be more interactive and create not only hermeneutic relations, but also alterity relations. This answer may also hint at the driving forces behind the cell phone's evolution from a voice-centric tool to a multimedia platform such as the iPhone, which displays a large variety of information types. During this evolution, more and more interactions with the cell phone are performed in relation to the screen. The screen has become a major component of the cell phone and the need to display data may explain a certain range of usages in which voice communications are not necessarily dominant. Alterity relations illuminate other usages.

By exploring Levinas' claims with respect to the face, one can acknowledge that the cell phone's screen has another form of faciality that we cannot ignore. As the screen becomes a quasi-face, the dialog with the cell phone becomes a face-to-quasi-face interaction. From a postphenomenological perspective, the cell phone technology is not only a quasi-other with which one maintains alterity relations. Following Ihde's notion of the "quasi-other," I suggest that the cell phone also has a quasi-face and, therefore, enhanced alterity.

It is important to remember that in alterity relations technologies are not mistakenly perceived as humans. Moreover, technologies that maintain alterity relations with their users do not fully replace humans nor are they an identical substitute. An example of an analysis in which technologies are considered a full replacement can be found in Sherry Turkle's *Alone Together*. She concentrates on the ways technology replaces human beings—as the object of romantic feelings, as parents and as children. But technologies do not replace humans. They do things that humans do in a different way: the wheel makes us move differently than our own legs; cell phones are there for us in a different way than our friends; the screen has a quasi-face. Moreover, the existence of the wheel does not render our legs redundant, although they are less used; cell phones do not make the institution of friendship obsolete, but rather change the modes of communications; a screen is not a "second face" but a quasi-face. Turkle further maintains: "I once described the computer as a second self, a mirror of mind. Now the metaphor no longer goes far enough. Our new devices provide space for the emergence of a new state of the self, itself, split between the screen and the physical real, wired into existence through technology."[61] My notion of quasi-face offers an alternative explanation, replacing Turkle's "schizoid" figure with a non-dystopian model of communication between users, other users and quasi-others.

My claim is that the screen is paradigmatic of a quasi-face. From a historical perspective, the quasi-face may explain why the cell phone's screen has been enriched with multimedia, so that it becomes closer to the face. It may also explain why numerous developers and manufacturers of consumer goods prefer to attach a screen to a variety of artifacts, including cell phones and refrigerators, in an effort to make them slightly more "human."

NOTES

1. With kind permission from Springer Science+Business Media: An earlier version was published in *Human Studies* (2014), 37, 299–316.

2. Emmanuel Levinas, *Totality and Infinity: An Essay on Exteriority* (Pittsburgh: Duquense University Press, 1995), 134.

3. Levinas, *Totality and Infinity*, 110.

4. Adriaan Theodoor Peperzak, *Beyond: The Philosophy of Emmanuel Levinas* (Evanston: Northwestern University Press, 1999), 134; Miguel De Beistegui, *The New Heidegger* (London and New York: Continuum, 2005), 20.

5. Levinas, *Totality and Infinity*, 110.

6. Ibid., 42–47, 77–78.

7. Ibid., 160.

8. Thus, intellectual property such as novels, movies or software code is excluded from their analyses of things. Like Heidegger, Levinas does not deal with Husserlian temporal objects, like music or written works, which require us to think in term of "consciousness of . . . ."

9. See the critic of Don Ihde, *Postphenomenology and Technoscience: The Peking University Lectures* (New York: State University of New York Press, 2009), 40.

10. Levinas, *Totality and Infinity*, 167.

11. As reflected in Heidegger's present-at-hand notion as well as in Deleuze and Guattari's notion of abstract machine. While Deleuze and Guattari refer also to modern technologies such as television, Levinas ignores television and even cinema, and remains bounded by the Heideggerian hammer and pen, or other traditional technologies such as clothing and dwelling. For Deleuze and Guattari, DTs and their users are entangled in such a way that the users are no longer users, but rather parts of larger systems. Their notion of machine allows us to imagine a future where the "users" will break the "larger machine" they compose with DTs. For Levinas, there is no need to break from technologies, because technologies help us live good lives.

12. Levinas, *Totality and Infinity*, 133, 110.

13. Ibid., 133.

14. Considered nowadays, it makes sense of the contemporary practice of setting a colorful wallpaper for the cell phone's display, without thinking about what it is good for.

15. Levinas, *Totality and Infinity*, 110.

16. Ibid., 233.

17. This basic approach to technology is also characteristic of postphenomenology which rejects both the technophobic and technophile discourses, and focuses on the possibilities that the combinations of human and technology open for us.

18. There is a vibrant discussion about whether Heidegger in "The Question Concerning Technology" is opening new horizon or looking back. For the former see Don Ihde, *Technics and Praxis: A Philosophy of Technology* (Dordrecht: Reidel Publishing Company, 1979); Graham Harman, *Tool-being: Heidegger and the Metaphysics of Objects* (Peru: Open Court Publishing, 2002); Scott Lash, *Critique of Information* (London: Sage, 2002); Don Ihde, *Heidegger's Technologies: Postphenomenological Perspectives* (New York: Fordham University Press, 2010);

Søren Riis, "Towards the Origin of Modern Technology: Reconfiguring Martin Heidegger's Thinking," *Continental Philosophy Review* 44 (2011): 103–117; for the latter see Andrew Feenberg, *Questioning Technology* (London and New York: Routledge, 1999); Robert C. Scharff and Val Dusek, *Philosophy of Technology: The Technological Condition, An Anthology* (Malden, Oxford, Victoria: Blackwell, 2003), 247; Amy E. Wendling, *Karl Marx on Technology and Alienation* (Basingstoke: Palgrave Macmillan, 2011).

19. Levinas, *Totality and Infinity*, 133.

20. A similar list of things necessary for living can be found in Karl Marx and Friedrich Engels' *German Ideology* (International Publishers Co, 1970): "life involves before everything else eating and drinking, a habitation, clothing and many other things" (48).

21. Heidegger refers to additional technologies such as the invisible and untouchable hydroelectric dam. These technologies do not require the hand, let alone the body of the user (here one should distinguish between an operator and a user). They are in the background.

22. For the role of the hand in the interface with a computer, see Stacy Irwin, "Technological Other/Quasi Other: Reflaction on Lived Experience," *Human Studies* 28 (2006): 453–467.

23. Levinas, *Totality and Infinity*, 167.

24. Ibid., 191.

25. Thus Deleuze and Guattari's Body without Organs is the next step in this direction of investigation and a further reduction to no organs at all.

26. Martin Heidegger, *What Is Called Thinking?*, trans. Glenn J. Gray and Fred D. Wieck (New York: Harper & Row, 1968); Cf. Jacques Derrida, "Geschlecht II: Heidegger's Hand," in *Deconstruction and Philosophy: The Texts of Jacques Derrida*, ed. John Sallis, trans. John P. Leavey Jr. (Chicago and London: University of Chicago Press, 1989), 198.

27. Frederick Engels, *Karl Marx and Friedrich Engles: Collected Works (Vol 25) - Friedrich Engels: Anti Duhring and Dialectics of Nature*, trans. Clemens Dutt (International Publishers, 1987), 330.

28. Levinas, *Totality and Infinity*, 159.

29. Ibid., 161.

30. Ibid., 160.

31. Ibid., 167.

32. Ibid., 177.

33. Ibid., 194; cf. Robert Eaglestone, *Ethical Criticism: Reading After Levinas* (Edinburgh: Edinburgh University Press, 1998), 117; Jacques Derrida, "Violence and Metaphysics: An Essay on the Thought of Emmanuel Levinas," in *Writing and Difference*, by Jacques Derrida, trans. Alan Bass (Chicago: University of Chicago, 1980), 122.

34. Bernhard Waldenfels, "Levinas and the Face of the Other," in *The Cambridge Companion to Levinas*, ed. Simon Critchley and Robert Bernasconi (Cambridge University Press, 2004), 63–81.

35. Derrida, "Violence and Metaphysics," 112.

36. Waldenfels, "Levinas and the Face," 63. See also Susan A. Handleman, *Fragments of Redemption: Jewish Thought and Literary Theory in Benjamin, Scholem, and Levinas* (Bloomington and Indianapolis: Indiana University Press, 1991), 209; Eaglestone, *Ethical Criticism*, 117.

37. Levinas, *Totality and Infinity*, 50.

38. Ibid., 202.

39. Ibid.; Handleman, *Fragments of Redemption*; Jacques Derrida, *Adieu to Emmanuel Levinas* (Stanford: Stanford University Press, 1999); Peperzak, *Beyond*; John Llewelyn, *Appositions of Jacques Derrida and Emmanuel Levinas* (Bloomington and Indianapolis: Indiana University Press, 2002).

40. Ibid., 66.

41. The words composing the discourse are a manifestation of the interaction with the other. Words are part of language. Levinas places importance on the role of language. He writes, "Language is not enacted within a consciousness; it comes to me from the other and reverberates in consciousness by putting it in question" (Levinas, *Totality and Infinity*, 204). Levinas adds, "language does not exteriorize a representation preexisting in me: it puts in common a world hitherto mine. Language *effectuates* the entry of things into a new ether in which they receive a name and become concepts. It is the first action over and above labor" (ibid, 174). Language, according to Levinas, is not a tool because it is abstract and because it is prior to labor. It conditions labor and is not part of labor. Language does not make interiority representable in an external spoken world, rather language is the discourse spoken that affects the self in being spoken. While Heidegger presupposes a language that already exists and into which one is born and with which one becomes human, Levinas maintains that language operates within us in a way more complex.

42. The visual representation, although unlike any other representation, is "[t]he manifestation of the face [which] is already discourse" (Levinas, *Totality and Infinity*, 66).

43. Levinas, *Totality and Infinity*, 67.

44. Handleman, *Fragments of Redemption*. Moreover, language participates in the thinking process. It not only "conditions the functioning of rational thought" (Levinas, *Totality and Infinity*, 204), but also "Language conditions thought" (ibid, ibid). This is in contrast to Heidegger, who thinks with the hand.

45. Levinas, *Totality and Infinity*, 67.

46. Ibid., 182.

47. Waldenfels, "Levinas and the Face," 65.

48. Ingrid Richardson, "Faces, Interfaces, Screens: Relational Ontologies of Framing, Attention and Distraction," *Transformation* 18 (The Face and Technology) (2010): http://www.transformationsjournal.org/journal/issue_18/article_05.shtml, 140; Waldenfels, "Levinas and the Face," 67–68; Peter Atterton, "Levinas and Our Moral Responsibility Toward Other Animals," *Inquiry* 54, no. 6 (2011): 622–649.

49. Levinas, *Totality and Infinity*, 160.

50. Here I use ontology in the sense proposed in STS and ANT, in which ontologies are always in the grammatical plural because, as Annemarie Mol asserts, "realities have become multiple" ("Ontological politics. A word and some questions,"

in *Actor Network Theory and After*, ed. John Law and John Hassard (Oxford and Malden: Blackwell Publishing, 1999), 75). As ontologies, the face is a condition for the human, for the animal and for the quasi-human. In other words, my claim is that there is more than one ontology for the face.

51. Don Ihde, *Technology and the Lifeworld: From Garden to Earth* (Bloomington and Indianapolis: Indiana University Press, 1990); *Peking University*. Language is the prime example of indirect mediation. When I look at a natural landscape, like the sea, my experience is mediated by words and concepts. The cell phone performs similar mediation via the built-in camera that mediates experiences. For many people, if something is not pictured, it did not happen to them. This explains why people take photos of the dishes they order in the restaurant and any move of a baby. Recently, this was exemplified during the election of the Pope in 2013 compared to that of 2005. In 2013, many of the people waiting in St. Peter's Square in Rome were raising their phones, laptops and tablets up in the air to take a picture of the new Pope's first speech; in 2005 only a few did so. See *Today Show*, https://www.facebook.com/today/posts/392824170816417 (accessed 23 April 2013).

52. Ihde, *Technology and the Lifeworld*, 100.

53. Ibid., 105.

54. Quasi-other should be carefully constructed in order to avoid anthropomorphism, through "A Phenomenologically Relativistic Analysis" (Ihde, *Technology and the Lifeworld*, 99).

55. Robrecht Vanderbeeken, "The Screen as an In-between," *Foundations of Science* 16 (2011): 245–257, 247.

56. http://outfit7.com/character/talking-tom/ (accessed March 20, 2013).

57. Levinas, *Totality and Infinity*, 204.

58. Ibid., 213. For these qualities of the human language, see also pp. 204 and 213.

59. Cf. Ibo van de Poel and Peter-Paul Verbeek, "Editorial: Ethics and Engineering Design," *Science Technology and Human Values* 31, no. 3 (2006): 223–236; Peter-Paul Verbeek, "Materializing Morality: Design Ethics and Technological Mediation," *Science Technology and Human Values* 31, no. 3 (2006): 361–380; Peter-Paul Verbeek, "Cultivating Humanity: Toward a Non-humanist Ethics of Technology," in *New Waves in Philosophy of Technology*, ed. Jan Kyrre Berg Olsen, Evan Selinger and Soren Riis (Basingstoke: Palgrave Macmillan, 2009), 241–266. There are other efforts to think of ethical robots, but most of those works discuss the role of the screen as an ethical focal point.

60. Levinas, *Totality and Infinity*, 201.

61. Sherry Turkle, *Alone Together: Why We Expect More from Technology and Less from Each Other* (Basic Books, 2012), 16.

Chapter 7

Third Invariant

Memory Prosthesis

The cell phone remembers thousands of phone numbers, messages from other people months and years back, the music we like, how many steps we have made every day (including the pathways), and much more. Many of these items cannot be memorized by most human beings without technological assistance. It used to be the technique of writing that did much of the memorization for past (Western) generations. Today the cell phone is unique in remembering all of these types of information in one device that can be carried everywhere. No wonder, then, that so many people take it with them everywhere, including to the toilet, and keep holding it even when they sit down and talk to friends. It functions as part of their body. It becomes a prosthesis. Because it remembers, it is a memory prosthesis.[1]

In this chapter I will examine the differences between memories of events and memories of processes. Then I will examine how memories of processes operate in relation to the body through a brief overview of enhancement, extension and prosthesis—three notions that denote the ability of technologies to assist humans. In the last section I will discuss the embodiment of the memory prosthesis in an effort to explain how the cell phone functions as a part of our body and mind.

THE MECHANIZATION OF MEMORY

Computing pioneer Vannevar Bush thought of human memory in terms of mechanics. During the 1930s and the 1940s he envisioned the Memex, a conceptual personal memory machine.[2] Conceived within the mechanical paradigm, it was intended to overcome the problems of decay, forgetfulness, and amnesia.[3] It was designed as a mechanical imitation of the human memory.

The machine, Bush thought, should mimic the biological way of thinking. That is, a way of thinking that ties two items together into an association. This approach to memory was novel at the time because it did not attempt to copy certain marks or freeze the fleeting like Freud's example of imprinting wax. Instead, it intended to mimic a process. The Memex was conceived at a time when significant efforts were being made to record events—through gramophone, radio, still cameras and cinema. Bush offered a unique approach to memory that can be regarded as a return to premodernity, in which the word memory indicated a process of the mind. Alternatively, this project can be conceptualized within the framework of the Industrial Revolution and in terms of factory machinery as recorded processes; in *Capital*, Marx provided comprehensive accounts of the ways in which the workers' gestures and knowledge were "recorded" into the factory machines. Some decades later, Fordism and Taylorism brought this recording to new heights. If the cell phone as an everyday technology records knowledge and gestures outside of the factory in an everyday context, then I will question whether there is a similarity in the effects it produces on humans.

Bernard Stiegler criticizes television and cinema (the "media industry") that record events, real or imaginary.[4] The media industry, he claims, "proletarianizes" consumers, just like the factory machines "proletarianized" workers. Stiegler adopts Karl Marx's notion of "proletarianization" which denotes the taking away of knowledge from the workers through implementing their know-how into machines. Stiegler claims this phenomenon is widespread and global. He writes, "The spread of industrial hypomnesic apparatuses causes our memories to pass into machines, in such a way that, for example, we no longer know the telephone numbers of those close to us—while the spread of spell checkers causes fear of the end of *orthographic consciousness* and of literary hypomnesic knowledge that goes with it . . ."[5] I find this critique of our inability to remember phone numbers to be somewhat dystopian. There is no value in the set of digits that represents a phone number. If human memory has to be practiced, as those preaching for memorizing phone numbers argue, then more significant strings of numbers should be memorized, such as the digital representation of the geometric π (Pi). If the critique concerns capitalistic exploitation, then another example should have been provided.

Leaving aside Stiegler's examples, the essence of his argument also invites a critical reading. In a nutshell, he argues against the media industry, in which the abilities to record and distribute certain events in a certain manner are controlled by a relatively small group that operates in order to increase its wealth. This is the meaning of what he terms "industrialisation of memory."[6] His argument is problematic in three ways: the focus on the event; the lack of reference to the changing landscapes of the media industry in the age of the Internet; and the emphasis on the role of selection what should be remembered.

Firstly, Stiegler bases his critique of the media industries on the event as the unit of memorization. For him "event-ization is memory's functioning."[7] This emphasis on the event leads Stiegler to overlook the memorization of modes-of-operation by DTs. Media technologies, in their digital form (i.e., the cell phone as media technology), record more than events—they also record processes. An analysis of contemporary technologies should refer to both aspects. For instance, the television records the mode-of-operation of a "couch potato," or, in other words, of the passive sedentary position. Examining the cell phone, the mode-of-operation that it records is more active, combining multiple media of talking, writing, running, taking pictures, etc. The cell phone records not only information and happenings but also modes-of-operation, processes, body gestures and habits. These may include asynchronous communication modes (in the forms of SMS, e-mails and updates via social networks) or chasing away boredom by listening to music and playing casual games.[8]

Secondly, unlike traditional radio, television and cinema technologies that are dominated by a relatively small group, DTs—from VCR, to the Internet and cell phones—have eroded the effectiveness of the selection of events by that group. In other words, the media industry and the factory are hierarchical organizations that have directed the development of certain technologies to support their hierarchies; those populating the lower levels (proletariat or consumers) cannot control the machines or the selection of memories to be recorded. This is the essence of the critique against twentieth century's capitalism since Adorno and Horkheimer. By contrast, DTs are mostly rhizomatic and enable much greater control and selection by all participants. They have the potential to offer an alternative way to that sought by capitalism.[9]

Thirdly, in contrast to Stiegler's description of automated selection of what should be remembered, the cell phone remembers everything including mistakes (such as pictures that were taken by the "swipe of the finger") and raw data (of constant monitoring of activities and body parameters). The element of selection that is at the heart of Stiegler's argument against the "industrialization of memory" is obsolete in the age of the cell phone. Everything is stored. Even what we delete frequently remains stored somewhere "in the cloud."

The cell phone's capacity to record events and processes runs across the historical variations and hence functions as an invariant. There is an additional type of memory that the cell phone enables: Because the cell phone can record memories as they are being created (e.g., traffic jams), such memories can be accessed in their "primary form," at the moment when they are produced.[10] These memories can be shared among people, even strangers.

Thus, access to other people's memory of a traffic jam while it is being recorded can help others avoid this obstacle and look for an alternative route. This memory impacts others and may change their future.[11]

There is a chasm between Bush's Memex and Stiegler's eventization. Although both attempt to fight amnesia, the Memex was designed to do so via the analog recording of human processes. DTs offer a third approach as they make irrelevant the problem of amnesia by simply recording everything without selection and without being subordinated to the hierarchies of the traditional media industry.[12] They function as an enhancement to the human memory.

ENHANCEMENT AND PROSTHESIS

By participating in the mechanization of memory, the cell phone functions as a cognitive enhancement. Anders Sandberg and Nick Bostrom review a variety of cognitive enhancements, ranging from chemical enhancements such as coffee and pills to medical machinery that sends pulses to the brain.[13] A parallel discourse uses the term "extension."[14] While the enhancement discussion focuses on the brain, the extension literature hardly refers to the body and focuses on the mind. Both discussions overlook the body that interacts with the technologies examined. This is where I find the concept of prosthesis appealing.

Prosthesis

The ability to remember is not unique to the cell phone. For Stiegler, this has been the essence of Technology since pre-history. He maintains that a tool reflects and at the same time establishes human memory. "A tool is, before anything else, memory," he claims.[15] As memory, Technology is the way for a human being to participate in a past that belongs to the individual although he or she never experienced it personally. It establishes the heritage and culture into which one is born and which shape one's identity. It is a memory recorded in the object, by the object and as an object. Consequently, "technics does not aid memory; it is memory."[16] Technologies—from tools, through machines, to DTs such as the cell phone—are memory recorders. They serve as media when they record events and as sociocultural registrars when they record modes-of-operation. The memory prosthesis lies at the triangulation of technology, memory and the human.

While the common meaning of prosthesis refers to an artifact that replaces a missing body part like a crutch or a peg-leg, Stiegler offers a new meaning for the word. He writes:

> A 'prosthesis' does not supplement something, does not replace what would have been there before it and would have been lost; it is added. . . . the prosthesis is not a mere extension of the human body; it is the constitution of this body qua 'human'. . . . It is not a 'means' for the human but its end.[17]

Technology is a prosthesis in the sense that it is an object-based memory of humans. As prostheses, technologies are the exteriorization of the human memory. By complementing the interiority of humans—and not through imitation as Bush thought—technology functions as a prosthesis. The prosthesis is not a simple copy of The Human but rather a transformative object. For instance, the invention of the wheel was not a recording of a memory of a certain type of movement but rather a new form of movement. Once the wheel was invented, the production of similar technological artifacts could be regarded as the externally recorded memory of what is Human. Vice versa, the exteriorization of "The Human" is the mnemonic function of technology. This double structure makes Technology un-dissociable from The Human. Stiegler's notion of prosthesis designates the co-constitution between humans and their technologies. However, due to his aim to critique consumerist society, he focuses on media technologies and event-recording capabilities, and hardly refers to the recording of modes-of-operation by contemporary technologies.

This notion of prosthesis offers a fresh view on human-technology relations, in sharp contrast to the common view that accentuate the "split."[18] The prosthesis is therefore coherent with the quasi-face, as described in the previous chapter.

Cognitive Prosthesis

The prosthesis has several modalities as it operates on humans in various ways. Helena De Preester maps three types of prosthesis, according to the capacities they extend:[19] (1) the motoric prosthesis that extends the limbs; (2) the perceptual prosthesis that extends the human sensorium; and (3) the cognitive prosthesis that enhances the mind. These types, she notes, are not distinct and technologies usually combine them. I find this mapping useful as an analytical arrangement of the extensions/enhancements/prostheses that technologies like the cell phone enable. Here I focus on the cognitive prosthesis.

The cognitive can be divided into various mental capacities. Sandberg and Bostrom offer a useful model of cognition as set of five processes that deal with information: acquisition (perception), selection (attention), representation (understanding), retention (memory) and use of the information to guide behavior (reasoning and coordination of motoric outputs). Combining this model with De Preester's mapping requires two clarifications. Firstly, although Sandberg and Bostrom's model can be criticized as referring to our mind in terms of computing—that is, taking the metaphor too literally—the model can help diffract the cognitive prosthesis and later systematically expand the prosthesis to domains beyond memory. Secondly, it should be noted that the five-process model does not fit as a whole into De Preester's three-type mapping: the first two processes are "perceptual" and hence can be accommodated by the second type of prosthesis, while the last three fit into the "cognitive prosthesis" category.

Sandberg and Bostrom's model is important for the analysis of cognitive prosthesis because it shows that the cognitive is not a single unified characteristic, but rather a collection of capacities. For example, it puts a question mark on Stiegler's attempts[20] to provide the same explanation for distinct phenomena—the broadcasting of events by cinema and television on one hand, and the cell phone's memorization capabilities on the other.

Within the various elements of cognition, De Preester limits her analysis of the cognitive prosthesis to the memory prosthesis. She cites Clark and Chalmers' well-known comparison between a brain-damaged person named "Otto" who has a notebook (an "outside") to record his memories and a healthy person named "Inga" who relies on her "biological" memory (an "inside"). She follows this example even though the juxtaposition of a disabled person (Otto) with a healthy one (Inga) sets us back to the limited notion of a prosthesis that replaces a defective organ, and can hardly accommodate a prosthesis that expands human capabilities. This example is also problematic from another aspect, as it deals with the embodiment not of the physical notebook but of its contents. Yet, Clark, Chalmers and De Preester do not discuss this distinction; they focus on the contents. For them there is no difference whether the content is stored in a notebook or on a cell phone. What matters to them is that it is external to the brain.[21]

Postphenomenology can enrich this analysis with its care for the phenomenological notion of experience. It is not sufficient to refer solely to the functionalist perspective as Clark and Chalmers do. Such a perspective leads to similar treatment of both internal and external memories (what I write in a notebook is equivalent to memorizing that information); a postphenomenological account reveals the differences in the experiences of the two types of memory. The experience is different when one remembers a phone number by heart and when one has the number stored on the handset. The first purports importance to each number memorized; the latter collects as many numbers as possible. Assigning priority to the memorization by heart reveals a somewhat technologically dystopian approach that prefers the human memory over the technological one, echoing antiquity's debate over writing as expressed in Plato's Phaedrus.

Clark and Chalmers maintain that there is no difference between the memorization of Otto and Inga. De Preester sees a difference in the experience of these memories and advocates the priority of the by-heart memorization through the notion of ownership. She focuses on the belief that a piece of information needs to be considered "mine" in order to construct memory prosthesis. In other words, the notebook should be owned in order for the memories contained in it to be considered as "mine." A cell phone should be owned, according to the rationale presented by De Preester, in order for the information contained in it to be considered as belonging to the user. She further asserts that the ownership of biologically stored information is "stronger." I disagree. Prosthesis is more than the strength of the consumerist relations we have with artifacts—the memory prosthesis is a priceless register of modes-of-operation, thoughts, wishes and events. The question of ownership of a medium should be separated from the question of ownership of contents, as evidenced in web-based services such as Flickr and Instagram. These services store people's pictures on remote servers that are owned by startups and corporations, not by the end-users who save their visual memories there. The mere storage of memories does not constitute ownership (in the mental sense) by the owner of the platform. In Clark and Chalmers' example, these web-based services are equivalent to a public notebook that is open to everybody, but the memories stored in it are still identified by the user as his own. Instead, I suggest, the mental attachment to memories should refer to their indexing logic, so that Otto must remember some important meta-data, such as the mere facts that (a) something was written, (b) where it was written and (c) how it was organized (sequentially? alphabetically? by topics?). What is important, I think, is the indexing of the memory and remembering where the information is stored in order to establish a link to past memories and therefore make them "mine." It is the acquaintance (with the content) and not the acquisition (of the medium) that matters.[22]

The importance of indexing in digital media can be traced back to the work of the interactive computing pioneer JCR Licklider who in the 1950s envisioned a computerized laboratory assistant that would arrange data and compile tables and graphs. A decade later, he led a related research project known as "Libraries of the Future."[23] In this project, the indexing of masses of information was the key challenge. This challenge remains relevant today. When an index is lost, the chances of finding the piece of information are slim. Losing information does not necessarily mean the information is no longer stored, but rather the loss of the index is sufficient for information to get lost. The enhancement/extension/prosthesis discussion teaches us that in order for a memory prosthesis to function, it must do more than store information. The memories should be organized in such a way that the user can find her memories in the ocean of data. This challenge has shifted from storage to indexing of masses of information. Now it is clear that the memory prosthesis is not a substitute for the human but rather is intended to augment and extend the human. It does not aim to imitate human memory; it is designed to maintain a symbiotic relationship with biological cognitive mechanisms.

In the rest of this chapter I sketch the guidelines for a new type of embodiment relations, which I term "augmented embodiment."

PROSTHESIS AND EMBODIMENT

The notion of prosthesis resonates with the postphenomenological notion of embodiment. Both refer to technological extensions of the human. Technologies that participate in embodiment relations are frequently referred to as prosthesis—that is, an external addition to the body. The embodiment relations can be motoric, as in the case of a peg-leg (which is the classical example of prosthesis in the commonly spoken language) or a car. It can also be sensory, as in the case of eyeglasses. The relations between cell phones and their users can be described in these terms: the cell phone is frequently held by the hand, and therefore becomes incorporated into the body schema. It extends the auditory sense by enabling the user to hear people from afar, much beyond human hearing capabilities. Embodiment relations as described by Ihde and others conflate the motoric and perceptual prostheses and widen their scope.

Embodiment relations, like prosthesis, are not limited to supplementing a missing organ (e.g., a peg-leg) or improving malfunction (as in the case of eyeglasses), but also encompass the meaning offered by Stiegler that adds capabilities which are not available to humans otherwise (e.g., "seeing" invisible radiation). Yet, all these senses of embodiment seem to be focused on the "brute" sense of exteriority of the body, and there are only a few references to the cognitive aspects of embodiment. A description of the cell phone in terms of "external" embodiment is not sufficient because this device extends the user's body scheme in a richer way that incorporates many other dimensions. The ability to find one's way even in a place never visited before changes the bodily comportment. Likewise, having all the photos, music, correspondence and contacts in one place certainly changes one's comportment.

In this section, I review some of the references to a certain embodiment which is beyond "exteriority." They expand the notion of embodiment relations to what I call "cognitive embodiment." It may sound to some readers like an oxymoron due to the apparent dominance of the body in embodiment, but it takes into account the difficulty of separating between mind and body, external and internal.

Body and Mind, Embodiment and Embrainment

Postphenomenology, like phenomenology, recognizes the inability to draw a clear line between body and mind, as well as between matter and thought, subjects and objects. From Merleau–Ponty to Ihde, embodiment is never viewed "mechanistically" and it is more than the mere attachment of an object to a human body.

Postphenomenology is not the only theory to interpret Merleau-Ponty in this way. Rosi Braidotti describes the dual influences of mind and body as "the embodiment of the mind and the embrainment of the body."[24] Embrainment complements embodiment. Together, these two notions describe the complex interrelations between mind and body.

Similar fusion is offered by N. Katherine Hayles as she coins the term "mindbody."[25] Mindbody reflects the difficulty in distinguishing between body and mind, even in the context of virtual reality. She rejects the modern glorification of consciousness and suggests instead a posthumanist approach according to which "consciousness for the posthuman ceases to be seen as the seat of identity and becomes instead an epiphenomenon, a late evolutionary add-on whose principal function is to narrate just-so stories that often have little to do with what is actually happening."[26] Identity, including memory, is not necessarily concentrated in the consciousness. Identity is distributed (and hence also memory), which results in the creation of the mindbody. Hayles provides a more complex interpretation of mindbody when she writes: "the individual person—or for that matter, the individual cyborg—is no longer the appropriate unit of analysis, if indeed it ever was. At issue now (and in the past) are distributed cultural cognitions embodied both in people and their technologies."[27] Identity can be found in the body as well as in the technologies one uses. Memory is no longer exclusively tied to consciousness but is distributed between the body and technological artifacts, such as cell phones. Consequently, postphenomenologists can pursue the posthumanist goal of "correcting an over-emphasis on consciousness to a more accurate view of cognition as embodied in human flesh and extended into the social and technological environment."[28] Embodiment, therefore, can be regarded as a fusion of interiority and exteriority whose ratios can change for various prosthesis types.

The notions of embrainment and mindbody set the ground for the cognitive embodiment relations. Some sub-types of postphenomenological embodiment relations can be classified as cognitive embodiment. One such sub-type is virtual embodiment in which the technology aims "to become the perfect simulacrum of full multisensory bodily action."[29] Here, technology aims to substitute the world that the user is experiencing. Virtual embodiment is an important permutation of cognitive embodiment because its major effect is on the mind. However, virtual embodiment requires a specific setting in which a user sits in front of a screen, which somewhat reduces the body to an inert and passive element. But my aim is to present the possibility of embodiment relations with a cell phone that functions as memory prosthesis. Embodiment with a cell phone does not require a sedentary motionless viewer, as is needed in cinemas or computer-based virtual reality systems. Nor is it limited to an imaginary virtual world, like the one provided by flight simulators. Being embodied with a cell phone means being immersed in the world and augmenting the experiences of this world with additional information.

Another permutation of cognitive embodiment is an embodiment that calls for hermeneutic relations in order for it to be successfully established. It is termed "implied embodiment."[30] Ihde's example is contemporary telescope that "display" unseen stars by spotting emissions such as radio, infrared and ultraviolet which are invisible to human perception.[31] It is an embodiment of the viewer and the telescope that extends the view beyond human sight. This viewing mode and its embodiment are based on some hermeneutic work in the translation of the inputs received from the telescope to human-readable star images. This embodiment is not performed directly with the telescope but with the hermeneutically mediated images. The technological mediation of the equipment is indispensable for deciphering the inputs. The parallel example for the cell phone would be the ability to "see" traffic jams on a possible route through a cell phone's navigation system, even if they are miles away. The cell phone's augmented reality applications require the device to be part of the user's body in the world, or in the reality she wishes to augment. Such an embodiment is based on the hermeneutic work of reading the additional information provided by the application. Implied embodiment is not suitable because it is somewhat limited to sensory extension. Because the cell phone extends not only the senses but also to cognitive faculties such as memory, a new type of postphenomenological embodiment relations that cover the case of memory prosthesis should be sought.

Augmented Embodiment

The embodiment I wish to explore here is a relation between a human and technology in which the human augments her mindbody so she can reach beyond human capacities, such as being able to read signs in a foreign language that one does not speak or detect traffic congestions several kilometers away.[32] Augmented reality as described in the fourth historical variation calls for a new conceptualization of embodiment in which technology is weaved into the user's body, perceptions, attention, understanding, memory and reasoning. I shall call this embodiment "augmented embodiment."

Augmented embodiment is in a sense a reply to an ancient technological utopia in which technologies enable humans to exceed their limitations.[33] But while most utopian dreams go after full embodiment, the augmented embodiment I suggest here operates as a partial addition to the human mindbody, as an addendum and not as a replacement or improvement of defective parts.

Like virtual embodiment, the effects of augmented embodiment are more cognitive than motoric or sensory. The interaction with the technology produces a certain worldview that is not necessarily physical. Like implied embodiment, augmented embodiment has a strong element of hermeneutic; the world has to be read through technology in order to make sense. Similarly to infrared-visible stars, a remote traffic jam needs to be technologically detected, transmitted and mediated in order to become part of drivers' experience and memories.

In other words, augmented embodiment offers a relation that is not limited to holding a notebook, but also includes embracing its content. Such an embodiment fuses mind and body.[34] It affects both the mind and the body that interacts with the technological artifact, including the memories stored in the artifact. In De Preester's terms, it covers all three types of prosthesis: the motoric, the perceptual and the cognitive.

If the option to detach determines whether a technology maintains embodiment relations with its user,[35] then the cell phone obviously maintains such relations. Yet, a sense of "undetachability" accompanies many everyday technologies. How many Americans are willing to give up their cars? Likewise, many cell phone users cannot imagine life without this technological artifact. As Ihde illustrates in the introduction to *Technology and the Lifeworld: From Garden to Earth*, we cannot imagine human life without clothing, cooking or dwelling technologies. They can be detached from the physical Body One but cannot be separated from the sociocultural Body Two.

Ciano Aydin reaches a similar conclusion as part of his post-phenomenological analysis of Clark and Chalmers' Extended Mind Theory. He writes, "Acknowledging that our thinking has an artifactual character means recognizing that external objects and technical artifacts, rather than being utilised by an inside world, have shaped and are continuously shaping the very fabric of our thinking, of what we take to be our 'inside world.' Not only are thoughts exosomatically embodied, but the specific physical characteristics of artifacts also activate new modes of thinking."[36] Aydin rejects the notion of extension and instead discusses the human technology relations in terms of "unfolding" and co-shaping. He formulates an alternative mind theory named the "Artifactual Mind Thesis." Aydin shows that not only is life impossible without technologies, so is thinking. Thinking is a major cognitive faculty that I did not discuss in this chapter. Instead I focused on another mind faculty, the memory, and examined how DTs like the cell phone participate in the storage and indexing of our memories.

EMBODIED MEMORY AND ARTIFACTUAL INTERIORITY

The prosthesis presented in this chapter has a meaning that exceeds the common definition of complementing a missing body part or correcting a malfunction. It is an addendum that augments given capabilities. When combined with memory, it denotes the capacity to remember more than what is considered average, common or biologically pre-given. The memory prosthesis merges the remembering of events with the recording of modes-of-operation. The cell phone—especially since the second historical variation in which it started functioning as media—urges us to combine the memory of usage processes with media memory.

There is a great affinity between the prosthesis and the postphenomenological embodiment relations because they both deal with the relations between a human body and technologies. Both notions can be expanded from the motoric and the sensory to the cognitive. Prosthesis and embodiment add new meanings to each other and call for the expansion of the postphenomenological relations to directions that are not covered by cyborg intentionality. The result is an augmented embodiment that is a relation of a detachable technology with a human mindbody, which forms a unit operating in the world.

Augmented embodiment assists in expanding the memory prosthesis to a broader range of memories—for example, memories-in-the-making and shared collective memories. This may explain why cell phones of the fourth historical variation access the real-time locations and memories-in-the-making of other people.

Embodiment frequently results in dependency of the human on the technological, even if the technological artifact is detachable. A peg-leg or a crutch, although detachable, is necessary for walking. The cell phone is necessary for many users to be in contact with others and with their own memories.

Future studies can be performed from three perspectives: that of the experiencing subject, that of the embodied object, and that of their combined point of view. The experiencing subject maintains augmented embodiment relations with the memory prosthesis (see Figure 7.1). The memory prosthesis records the user's modes-of-operation and events in which the user participates. As a unified functioning unit, the experiencing subject and the embodied memory prosthesis remember events and modes-of-operation. When combined, the integration between the human and the technological returns to the basic human wish that machines "become me."[37] Its parallel in the object could have been a wish of becoming human-like.

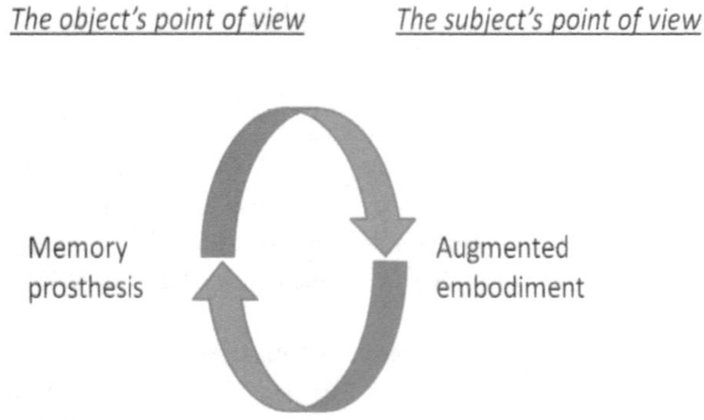

Figure 7.1 Memory Prosthesis and Augmented Embodiment. *Source*: Author.

This line of investigation aims at positioning the object on a plateau of its own that is not a reduced version of that of a subject. This positioning is the flat ontology. Flat ontology shifts the discussion from consciousness and mind to the relations between two elements in a system and the inherent inability to determine a clear, stable border between them. Flat ontology recognizes that not only the human develops and progresses in the process of co-constitution. Technology also evolves. This symmetry has (at least) two possible meanings: one confers a certain level of "consciousness" to artifacts. That is, an interiority that produces intentions; the other is more modest and suggests parallel processes of evolution. They are parallel and not similar because technology has its own evolutionary patterns, and any resemblance to human biological evolution is coincidental.[38] However, it seems to me that different processes may yield similar consequences, and so the rigorous historical determination does not touch upon the question of whether artifacts have an interiority filled with intentions, as suggested by the former meaning of the symmetry.

The question of artifactual interiority has a rich history dated to the late 1970s and mid-1980s. Langdon Winner asks in the title of an essay, "do artifacts have politics?" ANT's early works symmetrically referred to humans and nonhumans (e.g., Callon's fishermen and scallops) as "actants." Some twenty years later, Latour would call this sense of symmetry "absurd."[39] In between, the idea of symmetry has gained several followers.[40] One of them is Stiegler who explores in the first volume of *Technics and Time* the question of "who invents what." While everybody agrees that people invent technologies, Stiegler (following Leroi-Gourhan) questions whether the opposite is feasible. That is, if the Technical can invent the Human. In the form of Winner's provocative headline, Stiegler's question would be, "do artifacts have inventiveness?" Our question today can be "do artifacts have cognition?"

NOTES

1. There are additional types of prostheses, such as a social prosthesis, but this chapter deals with the memory prosthesis.

2. Vannevar Bush, "As We May Think," *The Atlantic Monthly*, July 1945. See Jay Hauben, *Vannevar Bush and JCR Licklider: Libraries of the Future* 1945–1965, http://www.columbia.edu/~hauben/lof-final05.doc. 2004; Belinda Barnet, "The Technical Evolution of Vannevar Bush's Memex," *Digital Humanities Quarterly* 2, no. 1 (2008).

3. For an analysis that denotes forgetfulness as an important (and positive) element of humanism see Bernard Stiegler, *Technics and Time, 1: The Fault of Epimetheus*, trans. Richard Beardsworth and George Collins (Stanford: Stanford University Press, 1998); *Technics and Time, 2: Disorientation*, trans. Stephen Barker (Stanford: Stanford University Press, 2009); Viktor Mayer-Schönberger, *Delete: The Virtue of Forgetting in the Digital Age* (Princeton: Princeton University Press, 2011). Moreover, forgetting today requires intention and effort as well as the intervention of the European Court of Justice, see David Streitfeld "European Court Lets Users Erase Records on Web," *The New York Times* , May 13, 2014 (http://www.nytimes.com/2014/05/14/technology/google-should-erase-web-links-to-some-personal-data-europes-highest-court-says.html, accessed 11 October 2014).

4. Stiegler, *Technics and Time, 2*.

5. Bernard Stiegler, *For a New Critique of Political Economy* (Polity, 2010), 30. This is not the first time that Stiegler laments lost capabilities such as remembering phone numbers by heart or driving a car (in the case of autonomous driving systems)—see Bernard Stiegler, "Anamnesis and Hypomnesis: The Memories of Desire," in *Technicity* ed. Louis Armand and Arthur Bradley (Prague: Litteraria Pragensia, 2006), 15–41. He further claims that this is a slippery slope in which we lose also "not only our know-how but also our savoir-vivre—and with it the little pleasures that make life worth living" (Stiegler, "Anamnesis," 17). Unlike previous contentions, here he does not provide any example and this contention remains declarative.

6. Stiegler's "industrialisation of memory" is his version of Adorno and Horkheimer's culture-industry induced reification of modern free and autonomous subjectivity. Today, according to Stiegler, we experience a clash of two "tendencies": one where the process of selecting which event enters our memory becomes industrialized; the other is that "industrialization ends in the development of [new] differant identities" (Stiegler, *Technics and Time 2*, 100). Because industrialization processes are "controlled by economically determined calculations" (ibid, ibid), the result is that "the question of the *who?* is one of a politics of memory encountering the resistance of an economic imperative" (ibid, ibid). This dystopian view was relevant when Stiegler wrote his first book in the trilogy of *Technics and Time* in 1994, when multichannel TV and Hollywood determined our cultural agenda by setting our memories and worldview. The emergence of personalized forms of contents (including user-generated content), coupled with the description I gave for the cellular phone, limits his critique to certain forms of television and cinema (cf. Bernard Stiegler, "The Carnival of the New Screen: From Hegemony to Isonomy," in *The YouTube Reader*, ed. Pelle Snickars and Patrick Vonderau (Stockholm: National

Library of Sweden, 2009), 40–59). For the transition from collective memories to connective memories see Andrew Hoskins, "Media, Memory, Metaphor: Remembering and the Connective Turn," *Parallax* 17, no. 4 (2011): 19–31; Anna Reading, "Seeing Red: A Political Economy of Digital Memory," *Media, Culture & Society* 36, no. 6 (2014): 748–760.

7. Stiegler, *Technics and Time 2*, 100.

8. In *The YouTube Reader*, Stiegler acknowledges the potential of Internet-based technologies to escape from the proletarianization scheme, but the reasons for this shift are not clear there.

9. These are rough sketches of the relations between humans, technologies and society, and the analysis should be more subtle due to the co-construction and co-constitution of humans and technologies.

10. This phenomenon is critically described by Stiegler as production of memories "in light time," leading to an inability "to distinguish an 'event' from its 'input' or its 'input from its 'reception' of reading" (Stiegler, *Technics and Time 2*, 116). The results are loss of sense of distance, loss of locality and loss of context. While this critique may be relevant for television as a technology for distributing news items, when mobile applications produce memories "in light time" the consequences can hardly be discussed in this all-negative ton.

11. Stiegler writes, "what memory tends to preserve is the future as incalculable" (Stiegler, *Technics and Time 2*, 128). Wrong again. Cell phones calculate Estimated Arrival Time (ETA); big data is the attempt to find the repeated patterns that will allow us to prepare to the future—weather conditions, traffic jams, etc. In the hands of large corporations it is a marketing tool that serves to shape consumers' preferences. In the hands of people, it is a tool to share the information their technologies gather and assist each other to prepare for tomorrow or for the next hour.

12. New structures may be identified, such as companies like Google and Facebook, but they function differently.

13. Anders Sandberg and Nick Bostrom, "Converging Cognitive Enhancements," *Annals of the New York Academy of Sciences* 1093, no. 1 (2006): 201–227.

14. See Andy Clark and David Chalmers, "The extended mind," *Analysis* 58, no. 1 (1998): 7–19; cf. Ciano Aydin, "The Artifactual Mind: Overcoming the 'inside–outside' dualism in the extended mind thesis and recognizing the technological dimension of cognition," *Phenomenology and the Cognitive Sciences* (2013): 1–22.

15. Stiegler, *Technics and Time 1*, 254. In later works he clarifies that "not every technics is a *mnemo*-technics" (Bernard Stiegler, "Our Ailing Educational Institutions," *Culture Machine* 5 (2003)). If memories are limited to events (that is—media), then Stiegler is right, but if memories also include modes-of-operation then it is difficult to think of technologies that are not memory capsules (or "mnemo-technics" in Stiegler's words). Technologies, I claim, were always part of the human memory (see Don Ihde, *Technology and the Lifeworld: From Garden to Earth* (Bloomington and Indianapolis: Indiana University Press, 1990)).

16. Stiegler, *Technics and Time 2*, 65.

17. Stiegler, *Technics and Time 1*, 152–153.

18. For example, Sherry Turkle, *Alone Together: Why We Expect More from Technology and Less from Each Other* (Basic Books, 2012).

19. Helena De Preester, "Technology and the Body: The (Im)Possibilities of Re-embodiment," *Foundations of Science* 16 (2011): 119–137.

20. Stiegler, *Technics and Time 2*; Stiegler, *For a New Critique.*

21. For an analysis of Clark and Chalmers' inability to overcome the internal-external dualism—see Aydin, "The Artifactual Mind"; Robert Clowes, "Thinking in the Cloud: The Cognitive Incorporation of Cloud-Based Technology", *Philosophy and Technology* 28 (2015), 261–296.

22. Clark and Chalmers discuss memories in the brain and in a notebook in terms of "access." It seems to me that such a terminology is based on a paradigm that considers our brain to be a large computer. While such a paradigm has led to certain advances in cognitive sciences, it has some limitations, and the understanding of human memory in these terms clearly marks the limits of the paradigm. We do not "consult" our brain and our memories are not "waiting to be accessed" (Clark and Chalmers, "The extended mind," 12). See also Marcel O'Gorman, "Taking Care of Digital Dementia," *CTheory* (http://www.ctheory.net/articles.aspx?id=740, accessed 24 May 2015).

23. JCR Licklider, *Libraries of the Future* (Cambridge: The MIT Press), 1965. Cf. Howard Rheingold, *Tools for Thought: The History and Future of Mind-Expanding Technology* (Cambridge: MIT Press, 2000); Hauben, *Vannevar Bush.*

24. Rosi Braidotti, *Nomadic Theory: The Portable Rosi Braidotti* (New York, Chichester, West Sussex: Columbia University Press, 2011).

25. N. Katherine Hayles, "Flesh and Metal: Reconfiguring the Mindbody in Virtual Environments," *Configurations,* 10 (2002): 297–320; N. Katherine Hayles, "Unfinished Work: From Cyborg to Cognisphere," *Theory Culture Society* 23, no. 7–8 (2006): 159–166.

26. Hayles, "Flesh and Metal," 319.

27. Hayles, "Unfinished Work," 160.

28. Ibid, 161.

29. Don Ihde, *Bodies in Technology* (Minneapolis, London: University of Minnesota Press, 2002), 7.

30. Don Ihde, *Experimental Phenomenology: Multistabilities (second edition)* (Albany: State University of New York Press, 2012), 131–143.

31. Peter-Paul Verbeek in "Cyborg intentionality: Rethinking the phenomenology of human–technology relations," *Phenomenology and Cognitive Science* 7 (2008): 387–395 also uses the contemporary telescope example when he discussed "composite intentionality." It is an extension of the postphenomenological relations in which human and technological intentionalities merge. He classifies it as an extension of the hermeneutic relations, that is—of the (representation of the) world. A similar type of embodiment can be detected in reading pictures of the surface of planets or deciphering of medical images (see Robert Rosenberger, "Perceiving other planets: Bodily experience, interpretation, and the Mars Orbiter camera," *Human Studies* 31 (2008), 63–75; Robert Rosenberger, "A Case Study in the Applied Philosophy of Imaging: The Synaptic Vesicle Debate," *Science Technology and Human Values* 36 no. 1 (2011), 6–32; Anette Forss, "Cells and the (imaginary) patient: the multistable

practitioner–technology–cell interface in the cytology laboratory," *Medicine, Health Care and Philosophy* 15, no. 3 (August 2012): 295–308).

32. A demonstration of these capacities can be found in a video clip that provides a futuristic outlook to an environment in which augmented reality applications become part of everyday life. It was prepared as a graduation project for Bezalel Academy of Arts by Eran May-raz and Daniel Lazo, "Sight," 2012, http://vimeo.com/46304267 (accessed 11 February 2013).

33. Ihde, *Experimental Phenomenology*, 131.

34. Cf. Verbeek's hybrid intentionality in his article "Cyborg intentionality."

35. Or cyborg relations, see Don Ihde, "Aging: I don't want to be a cyborg!" *Phenomenology and the Cognitive Sciences* 7, no. 3 (2008): 397–404; Verbeek, "Cyborg intentionality."

36. Aydin, "The artifactual mind," 2.

37. Ihde, *Technology and the Lifeworld*, 75.

38. George Bassala, *The Evolution of Technology* (New York: Cambridge University Press, 1988), 135–143; cf. Barnet, "the technical evolution" where she suggests that "retroactive and lateral 'transfer' of innovations is what distinguishes technical evolution from biological evolution, which is characterized by vertical transfer (parents to offspring)."

39. Bruno Latour, *Reassembling the Social: An Introduction to Actor-Network Theory* (Oxford and New York: Oxford University Press, 2005), 76.

40. Object-Oriented Ontology may also be regarded a part of this trend (see Ian Bogost, *Alien Phenomenology or What It's Like to be a Thing* (Minneapolis: University of Minnesota Press, 2010)), as well Clark and Chalmers' Extended Mind Theory.

Summary

Becoming-Mobile

BACK TO THE THINGS THEMSELVES

The research presented in this book is an analysis of everyday DTs. The case study chosen to represent DTs is that of the cell phone. By examining the cell phone, the research follows the Husserlian call "back to the things themselves." It follows a postphenomenological methodology of historical variations and forms a genealogy of the cell phone. There are four historical variations: the first concerns voice communications; the second introduces texts; the third marks the emergence of multimedia applications; and the fourth suggests sensory exploration as a base of augmented reality. The historical variations are comprised of "snapshots" of people, technologies and environments. Each variation examines how our being-in-the-world has been transformed in tandem with the increased use of cell phones, how cell phone technology has changed, and how the world has been modified. From the historical variations, three invariants were extracted: the wall-window, the quasi-face and the memory prosthesis.[1] The wall-window posits that the cell phone separates the user from her physical surroundings (the wall), and at the same time connects her to a remote space (the window). The remote space may be where the interlocutor resides or where information is stored (e.g., the Internet). The wall-window model can be further expanded toward media technologies and linked to the screen as a generic technology embedded in many technological artifacts. The central role of the screen in contemporary technologies led me to explore the question of why the cell phone got a screen in the first place. The second invariant is based on Levinas' notion of the face and Ihde's concept of the quasi-other. The screen, I argue, serves as a quasi-face, thereby enabling quasi-other technological artifacts. The screen keeps evolving toward a more accurate display in an endless effort to simulate a face. Parallel to the growth of the screen, the cell phone's memory has also been expanding. The third invariant is based on Stiegler's notion of prosthesis, where technology *is* the exteriorization of the human memory. For Stiegler, being human means transcending the biological memory of DNA and epigenetics into exteriorized memory inscribed in technological artifacts. I suggest expanding memory prosthesis to include memories as they are being produced with a future-looking orientation. Such an expansion is designed to contain the storage of telephone numbers to enable calls in the future as well as to access other people's routes in order to calculate the fastest route by bypassing traffic congestions.

One of the key challenges in conducting this work is locating the moments of constancy and a certain balance between "I," "technology" and "world." It is a relative-stability, which quickly turns into a multistability and even quasi-stability. Thinking about such situations in terms of being-in-the-world might dictate a certain stagnation, as it frequently leads one to focus on the stable rather than the changing.

BECOMING

Since Heidegger's *Being and Time*, we have been thinking of being. But being and being-in-the-world denote relatively static relations between a being and its environment.[2] In reality, these relations are never stable, they are constantly changing.[3] This state of change is the essence of becoming. Becoming means that there is no being that simply is. When Deleuze and Guattari employ the notion of becoming, they seek after what moves and remains unstable. Becoming is a *perpetuum mobile*.

Deleuze and Guattari demonstrate becoming and being in what they describe as the strategies of the bull and the ancient fish. The bull in the arena represents "being" in its attempts "to regain its associated milieu when danger appears" and to recapture "the turf it had chosen."[4] Obviously, these attempts are well expected and anticipated in the arena. The strategy of playing according to the rules even at the price of one's life is contrasted against that of the ancient fish that represents "becoming." Once the ancient fish realized that the seas have begun to dry, it made a courageous step and left the warm water to explore land. This exploration was accompanied by a long tedious process of "stand[ing] on its own legs" and "carrying water only on the inside, in the amniotic membranes protecting the embryo."[5] They add, "the animal is more a fleer than a fighter, but its flights are also conquests, creations."[6]

The ancient fish experiences the drying sea as a gradual change that slowly renders life unbearable and unsustainable. Instead of maintaining the same "being," as the bull does, the fish becomes and changes. Becoming is the acquisition of a new *modus operandi*. The change is not necessarily linear. Instead, it is a response to a stimulus that develops in various directions.

Deleuze and Guattari focus on becoming-animal and becoming-woman. While some of their examples, such as becoming-bird, may look esoteric, my claim is that we all go through multiple becomings. Becoming, I contend, is a general phenomenon. Growing-up is a becoming that each adult went through, accompanied by external changes in the body. But even as adults we keep changing. Becoming means there is no being that simply *is*. Becoming requires some effort and labor. Deleuze and Guattari note, "Even Jews must become-Jewish (it certainly takes more than a state)."[7] External visible signs or grandiose statements do not suffice. It is a process that requires an investment of energy.

Deleuze and Guattari are mostly interested in becoming-minoritarian, defined against the majority of Western male adult. They write, "There is no becoming-majoritarian, majority is never becoming."[8] And add, "You don't deviate from the majority unless there is a little detail that starts to swell and carries you off."[9] A change occurs as one differentiates himself or herself from the majority.[10] Rosi Braidotti interprets becoming-woman as overcoming and bypassing the traditional dichotomies of man-woman.[11] She recommends starting from a minority perspective so that diverse options are made available.[12] The research of the cell phone presented in this book follows Braidotti's guidelines. It attempts to overcome the dichotomy of fixed-mobile by starting the analysis from the "brick" cell phone rather than the landline telephone. This starting point allows me to conceive more options for the cell phone that exceed its functioning as a mere telephone. Here, it can be conceived as a writing device, an application accumulator or an augmented reality enabler.

There is another type of becoming in which the importance of the "internal" change is even greater: the becoming-molecular.[13] This may sound more familiar to non-Deleuzians because the combination of becoming and an adjective is commonly used in everyday speech. Within this category, Deleuze and Guattari coin the becoming-intense, becoming-expressive and becoming-cosmic, to name a few. Unlike the becoming-minoritarian, which refers to the target of the becoming (i.e., animal), the adjective in becoming-molecular denotes the type of change. My research project is about becoming-mobile, examining the traits rather than the target. Such a becoming has some characteristics that place it in proximity to co-constitution. I have identified five themes that characterize becoming: rhizome, molecular process, anti-memory, partial simulation, and transduction.

Rhizome

Becoming has no starting point or an end. It is open ended and grows in all directions. The rhizome is structured like the grass root, Deleuze and Guattari argue. It has a wild grid-like structure that resembles the street maps of

London or Beijing, rather than those of well-planned cities like New York or Paris.[14] The rhizome is the grid structure of the process of becoming. Because it has no center or a starting point, the rhizome should not be conceived in terms of progression or regression. The direction-less rhizomatic becoming may explain how the cell phone functions as a communication device for some, and a gaming device for others; a companion when walking in the dark and a memory prosthesis when taking pictures.

Deleuze and Guattari define becoming-as-rhizome by juxtaposing rhizome against "a classificatory or genealogical tree."[15] As a rhizome, becoming does not have a fixed arrangement like in a hierarchical model.[16] Moreover, becoming seeks to deconstruct the hierarchical (arborescent in their language) "organizing binary distributions within the dualism machines."[17] The Darwinian arborescent evolutionary vocabulary we frequently use when discussing a change should be replaced.[18] When Deleuze and Guattari determine that becoming is not an evolution, the rhizome is only an alternative, one option among others. Another option is the involution, which implies a process of change that is driven from the inside and that is recursive by its nature.[19] For example, internal system announcements in cell phones of the first historical variation can demonstrate how an internal "particle" drives involution. The textual aspects of these announcements may have marked the emergence of texting and the associated creative usage of acronyms and shortened words, and consequently pushed the cell phone from the first historical variation to the second. The seeds of change can be found inside in the form of built-in potentialities waiting for their wake-up call. Therefore, both rhizome and involution are efforts to bypass arborescent, hierarchical, and structural explanations, and open new understandings of evolution.

Molecular Process

Becoming is composed of small movements and tiny details, or in Deleuze and Guattari's terminology, it is molecular. The molecular is a fraction as opposed to the molar, which is a whole. It is a small, yet transformative, step done by the fish out of the ancient drying sea. The molecular is the level at which becoming occurs, because "the molecular has the capacity to make the elementary communicate with the cosmic."[20] At first, the claim that a small-scale molecular move can make a significant difference sounds illogical, but real-life examples can demonstrate how small changes may have a cosmic impact. A short text message sent from a demonstration to the outside world, or a picture uploaded from the cell phone to the Internet, are molecular moves that may create a molar effect.

Becoming does not necessarily mean that bodily organs change. It is a change in the way the body moves and in the way one holds one's body.

The example of becoming-animal may clarify this point. Deleuze and Guattari write, "Do not look for a resemblance or analogy to the animal, for this is becoming-animal in action, the production of the molecular animal."[21] Becoming-dog does not transform the human body to have four legs and a tail. Yet, Deleuze and Guattari add, "becoming does not occur in the imagination, even when the imagination reaches the highest cosmic or dynamic level . . . Becomings-animal are neither dreams nor fantasies. They are perfectly real."[22] Even if it loosely resembles an external entity (e.g., an animal), the process of becoming can be spotted, noticed and inspected. Between a change in the body and a virtual change in the imagination, becoming is a change in the mode-of-operation. In this process of change, the inside "takes possession of the body's interior, enters its cavities, and fills its volumes instead of making a face for it."[23] Becoming is performed on the interiority of the body, and might change the exteriority as a side effect.

Anti-memory

Becoming is a future-looking process, yet Deleuze and Guattari assert that becoming is an antimemory.[24] Becoming does not return to a fixed past, but instead orients toward an unknown future and a redefined past. Therefore, "history is made only by those who oppose history (not by those who insert themselves into it, or even reshape it)."[25] To start a revolution, the revolutionist needs to break away from the past just as the ancient fish had to forget the sea. In the technological realm, we have seen that the argument that cell phones cannot be used to access the Internet had to be rejected in order to develop a cell phone of the third historical variation. In order to become, one needs to forget what one cannot become. In order to use a cell phone as an application accumulator, one has to forget it is a "phone" and instead use it as a device to access the Internet. This is the meaning of "histories of the future."

Partial Simulation

Deleuze and Guattari warn their readers that becoming does not produce an identical copy and not even a resemblance, similarity or analogy. Nor is it an imitation process.[26] Instead, becoming is more like infection and contagion.[27] Braidotti characterizes becoming as "affinity and symbiosis between adjacent particles."[28] The "becomer" needs to be attracted to or feel sympathy for the target of the becoming so that an exchange can be established between the "becomer" and the object of becoming. Similarly, Keith Ansell Pearson coins the neologism viroid to imply a relation of exchange of fractions of code performed by viruses.[29] Such viral reproduction schemes do not necessarily create an identical offspring. In becoming, the result of the process is not identical to the original or the model, but instead borrows some attributes from it in order to partially simulate its mode-of-operation.

In becoming-animal, the change is not a creation of an actual copy of a specific animal. Rather, it is a process of change that simulates certain aspects of behavior, thinking, and desiring. It is a partial simulation. The cell phone does not imitate our way of remembering, but we become accustomed to the way it produces memories and how they are stored—for example, pictures separated from texts. We maintain a symbiosis with the cell phone that holds our memories. The cell phone is our memory prosthesis, and a code-of-memory is shared between us and this technological device.

Transduction

Deleuze and Guattari observe, "Becoming is always double, that which one becomes becomes no less than the one that becomes."[30] They provide several examples for the transductivity of the becoming process: "The human musician is deterritorialized in the bird, but it is a bird that is itself deterritorialized, 'transfigured,' a celestial bird that has just as much of a becoming as that which becomes with it. Captain Ahab is engaged in an irresistible becoming-whale with Moby-Dick; but the animal, Moby-Dick, must simultaneously become an unbearable pure whiteness, a shimmering pure white wall . . ."[31] Becoming influences both parties. The transformation occurs not only in the one who becomes but also in the target of the becoming, even though these transformations are not not-necessarily reciprocal, simultaneous or coherent. Becoming is a transductive process in which a transformation takes place in more than one element.

Etymologically, the word transduction originates from the Latin verb *tranducere* which means to change over or convert. If *in*duction is the process of deriving general principles from particular facts, and *de*duction goes in the opposite direction, that is, from the general claim to the particular cases, then *trans*duction seems to aim at catching both ends of the general and the particular.

Gilbert Simondon retains the sense of reciprocity and defines transduction as:

> A process—be it physical, biological, mental or social—in which an activity gradually sets itself in motion, propagating within a given area, through a structuration of the different zones of the area over which it operates. Each region of the structure that is constituted in this way then serves to constitute the next one to such an extent that at the very time this structuration is effected there is a progressive modification taking place in tandem with it.[32]

Simondon focuses on a single system composed of two parts and delineates the relations between them.[33] He exemplifies transduction with a crystal that develops in its solution-liquid. A change in the liquid's balance forms a grain of a crystal, and the grain of the crystal produces imbalance in the liquid. This process is not limited to chemistry; double becoming is well known in biology where insects and flowers match reproduction systems, such as that of the red clover and the bee, and the orchid and the wasp. In these examples each party retains its own identity. Yet, as they get closer and live in symbiosis, they influence each other. They adapt to each other's needs and abilities. As a transductive process, becoming emphasizes the simultaneous changes of a dualistic ecosystem.

Transduction contributes to the conceptualization of the change process by revealing the directions in which becoming develops. Simondon details,

> The ultimate terms at which the transductive process finally arrives do not pre-exist this process. Its dynamism derives from the primitive tension of the heterogeneous being's system, which moves out of step with itself and develops further dimensions upon which it bases its structure. It does not derive from a tension between the terms that will be found and registered at the furthest edges of transduction.[34]

A transductive process cannot be well planned and well designed. It is more exploratory and adventurous; it cannot be otherwise. The results are always surprising. Indeed, Simondon clarifies, "clearly, transduction cannot be presented as a logical procedure terminating in a conclusive proof."[35] Becoming, like co-constitution, transcends the intentions of engineers and designers. The meeting of technologies and humans in the world results in surprising phenomenona like the "designer fallacy" and the decentering of the inventor.[36]

By adopting Simondon's transduction, Deleuze and Guattari seem to prefer the model of involution over evolution, which is based on a change that happens from within the system and to some extent ignores external influences. The crystal-in-liquid example is a fairly simple ecosystem. Technologies, cell phones included, pose a more complex system. The prefix *co* of co-constitution denotes the simultaneousness and interrelatedness of the change processes that complex systems go through. Co-constitution denotes the changes in humans as much as in their technologies, and in technologies as much as in the world.

The Process of Becoming

Becoming is a rhizomatic, molecular, future-looking, viral and transductive process. Deleuze and Guattari argue that becoming is a process between two beings that "is neither one nor two, nor the relation of the two; it is the in-between . . ."[37] While the analysis of becoming is often concentrated on two elements, the becoming I employ in this work explores the transformative processes of "I," "technology" and "world." It emphasizes the dependency of the various components on each other in the process of change. Like co-constitution, co-construction and co-shaping, what is important is the interdependency of the components of the scheme. While Deleuze and Guattari offer the example of becoming that refers to transformation of a subject or an object, the concept of becoming that is used here accentuates the claim that change occurs transductively. Consequently, the becoming of "I," "technology" and "world" is not comprised of separate changes that necessarily happen at once and in the same manner. It is a becoming in which each component shapes and is shaped by the others, a transductive process that erodes the borders between them.

When I store my memories on a cell phone—e.g., phone numbers and pictures—the border between the human and the technological is not a clear-cut line. Memories can reside either in my head, in the handset or in a cloud on the Internet, and each storage mode has its own implications.[38] However, content-wise, these memories are always mine. Memories travel between humans and technologies, and the border in-between is not a simple area; it is a block "essentially mobile, never in equilibrium."[39] Memories flow within the block, so that the system of me, my cell phone and my memories is never stable.

BECOMING-MOBILE

The cell phone opens new modes of mobility in which the user's body can remain stationary while voice, text and data flow to and from the handset—a dynamic that erodes the well-established distinctions of mobile-immobile, connected-separated and active-passive.[40] This is a partial mobility; it produces partial movement since only certain aspects of the user are transported. In Negroponte's terms, it involves the movements of atoms and bits in a variety of combinations. It consists of only moving bits when the user is stationary, or a synchronization of bits and atoms in the case of augmented reality. I propose the notion of becoming-mobile to denote the movement of atoms and/or bits. In the next sub sections I examine the concepts of virtual mobility (as developed in New Mobility Studies) and of Deleuzian nomadism in order to explore what becoming-mobile is and what it can be. After assessing the contribution of each of these concepts to becoming-mobile, I detail how the five traits of becoming are relevant for becoming-mobile and how becoming-mobile functions as an invariant of the invariants. Lastly I show that we have always-already become-mobile.

Virtual Mobility

The partial movement enabled by the cell phone has recently attracted the attention of historians of technology. This type of research can be found in New Mobility Studies, which not only examine the history of transportation but also theorize movement in general.[41] Movement and flows are explored in a transdisciplinary way that extends the research to what Heike Weber terms "virtual mobility."[42] Virtual mobility allows us to understand the mobilization of parts of us, as in the case of telecommunications technologies that transport our voice across distances. Weber describes the cell phone as "an emblem of mobility and freedom" for combining traveling with communication.[43] In this context, the cell phone is used primarily for connecting to other people, and mainly covers the first and second historical variations.[44]

However, there are certain effects, such as rhizome and simulation, that the concept of virtual mobility does not take into consideration. Additionally, due to its historical and cultural research origins, virtual mobility does not provide an account of the experience of a user who is partially transported when voice, text or memories travel as bits while atoms remain stationary. Lastly, the analysis is limited to the first two variations, and does not conceptualize the rhizomatic experience of transporting bits and atoms when the user moves while being accompanied by augmented reality applications. Virtual mobility, with its emphasis on the partiality of the early historical variations, should therefore be expanded in two orthogonal directions: one toward the more recent historical variations and the other toward the additional characteristics of becoming beyond partiality.

Nomadism

Becoming-mobile (like virtual mobility) aims at framing the oxymoron of movement while being stationary. Deleuze and Guattari think of this oxymoron in terms of nomadism,[45] a term that describes "motionless voyage"[46] and "station as a process."[47] Beyond the shared oxymoron, becoming and nomadism are intimately related so that "the nomad exists only in becoming and in interaction."[48] To be nomad is to become through an interaction.

Unlike the concept of virtual mobility, which was developed for studying the historical-cultural effects of movement in a given geographical location, nomadism refers to entities that are constantly moving regardless of their geography. Indeed, the historical variations of the cell phone can be spotted across countries and continents. They are not based on inductive research but on a phenomenological generalization. Furthermore, virtual mobility is interested in the historical changes that societies go through, while nomadism does not presume a stable or durable entity. Rather, it attempts to understand the structures of restless groups and individuals.

Nomadism is primarily employed to express the movements of people and societies. Eugene Holland characterizes a nomad society as "placeless, itinerant tribes people," that is contrasted with "groups whose organization is immanent to the relations composing them."[49] The nomads, Holland maintains, operate like an improvisational jazz band, a manner which is substantially different from a strictly organized symphonic orchestra. If a nomad movement is about jazzy adjustments that make up a loose composition, then virtual mobility is about a perceptible symphonic social order that is transformed.

Nomadism can be extended from people to the space they inhabit. Claire Colebrook refers to the nomadic space "not as a space with intrinsic properties that then determine relations (in the way chess pieces determine how movements might be enacted), but as a space with extrinsic properties; the space is produced from the movements that then give that space its peculiar quality (just as in the game of Go the pieces are not coded as kings or queens but enter into relations that produce a field of hierarchies)."[50] Nomadic space is created by an unregulated movement, while a virtually mobile space assumes given positions from which and to which one can move. The cell phone can demonstrate how a nomadic space functions. When a phone call is performed, a new space "transpires" where a user and an interlocutor "meet"—this is the first historical variation. While this new space is brought into existence by all phones, including fixed-line telephones, the cell phone enables access while one is on the move (provided there is cellular coverage, of course). Once text becomes part of the cell phone, as in the second variation, a more abstract space is created. That space is asynchronous; it provides an exchange of text messages (and of e-mails as in the third variation), which does not necessarily take place at the same time. This space does not exist prior to the call or the sent text message. Instead, this space "transpires" by the call or the sent text message. The cell phone's call or text message generates a nomadic space that is produced while being on the move. It is a virtual space, which is named cellspace in the fourth historical variation.

Nomadism does not only characterize societies and spaces. It also designates thinking structures of individuals[51] and certain modes of scientific investigation.[52] Becoming-mobile aims at expanding nomadism to experiences of mobility, to the interaction between users and cell phone in the world, and in the intersection of I-technology-world. While nomadism is based on a nomad who has no geographical home base, the cell phone user has a dwelling from which he or she uses the cell phone in a nomadic fashion. What is unique to the cell phone user is that he or she are both virtually mobile and nomad, or in other words—partially nomad. The cell phone's user becomes-mobile.

The Process of Becoming-Mobile

I term the process of the transformations of humans, their cell phones and the environment as becoming-mobile. It is a virtual mobility as conceptualized by New Mobility Studies and has a jazzy nomad spirit like the Deleuzian nomadism. Unlike physical mobility that can be described as a two- or three-dimensional movement within certain geography, becoming-mobile is about the capacity to move from a physical space to a virtual or augmented space, back and forth, and maintain a position in more than one space at a given time. Not limited to home, office, or any other fixed point in space, becoming-mobile is based on partial spaces—auditory, visual or virtual—and it can happen everywhere. Moreover, it is not necessarily a physical movement in space, though it can accompany such movement. It can encompass various forms and formats, including those of stationary bodies. But it does not necessarily require the movement of an individual from one space to another. Becoming-mobile implies an experience of virtual mobility in a nomadic space.

Becoming-mobile corresponds to becoming's five characteristics:

1. Becoming-mobile is a rhizomatic movement. One moves in and between real, virtual and augmented spaces. There is no starting point or a pre-given destination. No space is superior to another and the order of these spaces is not pre-arranged.
2. Becoming-mobile is molecular in two complementary aspects. It is molecular in the electronic milieu where the user and the world are transcoded into digital signals, much like the production of the bird's voices in Hitchcock's *The Birds*. "When Hitchcock does birds, he does not reproduce bird calls, he produces an electronic sound. . ."[53] Likewise, the voices of the user and the interlocutor are not sent as is over the cellular network, but rather their representations are sent as series of bits, coded and decoded.[54] The same applies for pictures and videos. While becoming-molecular takes place in the real world of atoms, transcoding into digital takes place in the world of bits. Becoming-mobile is therefore more about becoming-electronic or becoming-digital, terms which may emphasize the importance of the production of an electronic representation of sound, image and text.

   The other aspect of molecularity is expressed in the tiny molecular actions, gestures and operations that comprise the becoming-mobile. In becoming-mobile, the user's environment changes even at the sound of a ringtone; the sound of an incoming call transforms the public space into a private and intimate space of the user. The "privatization" of the public space into the user's personal space is instantaneous. A small-scale action such as answering a phone call can "undermine the great molar powers of family, career, and conjugality."[55]

3. Becoming-mobile is anti-memory; it encourages the user to let go of prior usages and reproduce them in new constellations. By texting a message on the cell phone, the user "forgets" how to use a telephone and reconstructs it as a writing machine.
4. Becoming-mobile is a partial movement as depicted in the concept of virtual mobility. The partiality extends to other dimensions such as the adoption of certain modes-of-operations by way of a reproduction of partial codes and memories, like a virus that leaves some genetic code with its host and adopts others' codes. The user changes his or her behavior, and in turn the cell phone changes through a set of iterations of the user's memories and habits. Becoming-mobile does not mean that the user becomes the object of his or her becoming. According to Deleuze and Guattari, a child who becomes-animal does not transform into the animal; the child is partially transformed. She converts certain elements in her behavior and retains others. Similarly, users of the third historical variation become accustomed to splitting their memories by media, storing pictures in one space, and exchanging texts before, during, and after taking the picture in another space. In becoming-mobile, a partial process of transformation takes place. There is no full transformative process of identity. Becoming is a process of acquiring new modes-of-operation.
5. Becoming-mobile is a transductive process in which humans, technologies and environments transform and co-shape each other in surprising and not-necessarily reciprocal, simultaneous or coherent ways.

Becoming-mobile is a new mode of mobility that alters the user's relations to the world. In postphenomenological terms, it encompasses the embodiment relations one maintains with a cell phone while moving bits or atoms or both. It offers new hermeneutic relations as the user relies on information enabled by augmented reality applications. Becoming-mobile provides new alterity relations by allowing new modes of being-with-others.[56] It also provides a new way of being-with-oneself and transforms modes of boredom and interest.

Becoming-Mobile as an Invariant of Invariants

Becoming-mobile can function as an invariant of the invariants. This construct is a "doubling," in that a set of invariants can be grouped and gain new meaning under an invariant of invariants. In the introductory chapter I described the invariants as warps that compose the carpet's foundation, yet remain hidden underneath the wefts. The invariant of the invariants can be considered to be the frame into which the invariants are "stringed" (see Figure S.1).

As a frame, it unites the historical variations and the invariants. As an invariant of invariants, it does not function as an essence or a capacity, but rather as a principle.[57] It is a common thread that is weaved throughout the wall-window, the quasi-face and the memory prosthesis invariants.

Each invariant becomes mobile as it remains stable through a transformation that is rhizomatic, molecular, anti-memory, partial and transductive. I shall demonstrate this claim with one of the extreme intersections of invariants and becoming—the memory prosthesis as anti-memory. The cell phone can function as a memory prosthesis if and only if the user "overcomes" some modes-of-operation and opens a horizon for a new type of memory. In the first historical variation, one had to leave behind the habit of remembering phone numbers and allow the handset to memorize the important phone numbers of home, office and relatives. In the second variation, voice communication had to be put aside to not only allow texting but also non-communicative short texts, such as meeting descriptions in the calendar or a task in the to-do list. In the third variation, communication gave way to applications, and everything was reduced to an "app," including a flashlight, a compass and a mosquito chaser. Communication (especially as a major characteristic of a cell-*phone* or a smart-*phone*) has to be withdrawn in order to use these applications. The example of memory prosthesis as anti-memory demonstrates how becoming-mobile is an active thread of the invariants. Becoming-mobile is an invariant of invariants that is applicable to additional invariants such as the companion à-la-Haraway, which was not described in this work.

There is a tension between the terms "becoming," which denotes movement and dynamism, and "invariant of invariants," which denotes a certain stability and even inactivity. The reader should not replace the notion with its content. While the frame is constant by definition, its content can be a reference to movement and change. Another tension is related to essentialism, according to which becoming-mobile—due to its widespread character—can be understood as an essence of the cell phone's technology. But as an

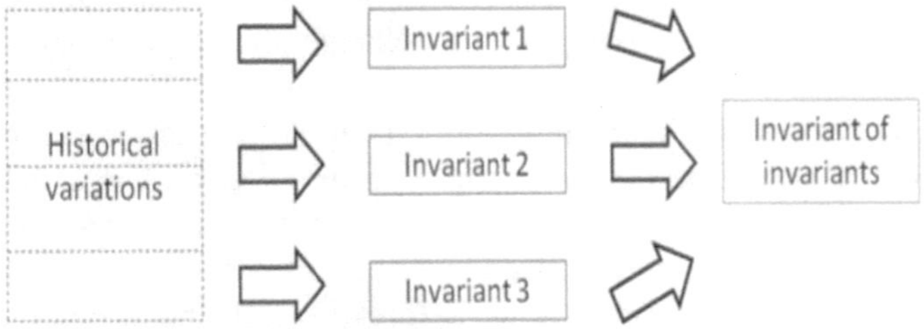

Figure S.1 From Historical Variations to Invariants. *Source*: Author.

invariant of invariants it preserves a specific characteristic of the invariant—that invariants are plural, which distinguishes invariants from the single Heidegerrian essence. It means that more than one frame can fit a certain set of warp and weft.

Moreover, becoming-mobile serves to describe the "how" more than the "what" of cell phone technologies and their relations with their users and the world. This direction of research is different than that pursued, for example, by Heidegger in *Being and Time*, where he concerns himself with the question of the meaning of being, or in "The Question Concerning Technology" where he concerns himself with the question of the meaning of technology. This work is not about the meaning of a cell phone; it is about the transformations in the relations between people, technologies and the world. Eventually this question may lead to more ontological questions, but these are beyond the scope of this project.

At the beginning of this chapter I argued that co-constitution is the mechanism that operates in each historical variation. I claimed that it is the moving power behind the "progress" from one variation to the next, so that co-constitution denotes the common processes of change that occur between humans, their technologies and their environment. After studying becoming and becoming-mobile, I would like to slightly rephrase that statement and argue that becoming-mobile can be extended from the category of invariants to that of variations to describe the mechanism that moves users, technology and world from one variation to the next. This movement is rhizomatic in that it has no predetermined or fixed direction (in the sense of progress). It is molecular since it is expressed in minute details, such as harbingers of text in the first historical variation as evidenced in the handset's internal announcements (e.g., "totAL" written on a LED screen which was designed to present digits only). Anti-memory assists in the shift from one variation to the next because the user has to forget some functionality in order to enable new usages. Movement to the next variation is always partial and a clear border line between the variations cannot be sketched (for example, iPhone was not the first "smartphone" with applications and its launch cannot mark the beginning of the third historical variation, although many tend to mark it as such). Lastly, the movement between the variations is transductive because each element of the I-technology-world scheme affects the others in a non-simultaneous noncoherent way.

Once we realize that becoming-mobile is applicable to the invariants and is the moving power between one variation to the next, a new understanding that we are always-already becoming-mobile arises. As shown for the wall-window invariant, the separation capabilities and the opening of an alternative space had operated long before the cell phone emerged. They can be found in written texts, in paintings, in story-telling and many other technologies and technics. Likewise, the becoming-mobile of the quasi-face operated in ancient history when people carried small figures made of clay, stone, wood or gold denoting gods equipped with a certain type of quasi-faces. These ancient ancestors of ours have been also always-already carrying their memory prostheses in the form of tools that encapsulated various modes-of-operation. We are always-already becoming-mobile.

Becoming-mobile may conceptualize the evolution (and involution) of additional technologies beyond the cell phone because it is a general frame, or an invariant of invariants. It is as general as embodiment relations that cannot be limited to a specific technology or to a certain era. Becoming-mobile conceptualizes the multiple movements of atoms and bits in several spaces, whether simultaneously or not, whether coherent or not, whether dependent or independent. It bears nomadic logic of smooth movement in all directions combining the virtual with the real.

NOTES

1. Additional frameworks can be employed as invariants. For example, Donna Haraway's companion species can conceptualize the co-dependencies between people and cell phones, equivalent to the co-dependencies between humans and dogs ("Cyborgs to Companion Species," in *Chasing Technoscience: Matrix for Materiality*, ed. Don Ihde and Evan Selinger (Bloomington: Indiana University Press, 2003)). And there are more invariants that can be sought.

2. For contrasting processes (or becoming) and things, see Dorothea Olkowski, "Flows of Desire and the Body-Becoming," in *Becomings: Explorations in Time, Memory, and Futures*, ed. Elizabeth Grosz (Ithaca: Cornell University Press, 1999), 100.

3. William E. Connolly provides a nuanced view of the constant element of Deleuzian becoming. He rejects an interpretation according to which "everything is always in flux," and suggests instead an oscillation "between periods of relative arrest and . . . heightened imbalance and change, followed again by new stabilizations" (William E. Connolly, *A World of Becoming* (Durham and London: Duke University Press, 2011), 44). My take is that real-life cases prove how difficult it is to find a period of stabilization with no change. Changes always happen, and the question is of frequency and intensity.

4. Gilles Deleuze and Felix Guattari, *A Thousand Plateaus: Capitalism and Schizophrenia,* trans. Brian Massumi (Minneapolis, London: University of Minnesota Press, 1987), 55.

5. Deleuze and Guattari, *A Thousand Plateaus*, 55.

6. Ibid., 55.

7. Ibid., 291. Jews, like woman, are considered by Deleuze and Guattari as minorities.

8. Ibid., 106.

9. Deleuze and Guattari, *A Thousand Plateaus*, 292.

10. Yet, a minority is a state-of-mind, because "A minority can be small in number; but it can also be the largest in number, constitute an absolute, indefinite majority" (469). For example, Occupy Wall Street has shown how the 99% became a minority in the Capitalist Western countries.

11. Rosi Braidotti, "Becoming Woman: Or Sexual Difference Revisited," *Theory Culture Society* 20, no. 3 (2003): 43.

12. Braidotti, "Becoming Woman."

13. For Braidotti, however, they are the different sides of the same phenomena (see "Becoming Woman"). Instead she differentiates between a becoming that "operate[s] a much-needed dislodgement of dominant subject positions (masculinity, hetero-sexuality, whiteness, gerontocracy, Euro-centrism in the imperialist mode)," which is equivalent to the becoming-minoritatrian, and a becoming which "mark[s] instead the conditions for the affirmation of new subject positions and thus lay the foundations for possible futures" (55), which is the result of the becoming-molecular.

14. This is, by the way, the essence of ANT's notion of network; Bruno Latour, "On Recalling ANT," in *Actor Network Theory and After*, ed. John Law and John Hassard (Oxford and Malden: Blackwell Publishing, 1999), 15–25.

15. Deleuze and Guattari, *A Thousand Plateaus*, 239.

16. Lyat Friedman, "Thousand Plateaus: An anti Nietzscheian act," *Theory and Criticism* 19 (2001): 231–239.

17. Deleuze and Guattari, *A Thousand Plateaus*, 292.

18. Belinda Barnet ("Technical Machines and evolution," *CTheory* 2004) explains the differences between Darwinian and technological evolution, and specifically describes the inability to draw a hierarchical evolutionary diagram for a collection of cornets.

19. Deleuze and Guattari, *A Thousand Plateaus*, 238. It is known in biology that major evolutionary phases, such as the emergence of a new phylum, start when the genes "open up" and express differently. Keith Ansell Pearson notes that the difference between humans' and chimpanzees' DNA is minimal, and most of the differences are in the spatial organization and folding of the genetic material (Keith Ansell Pearson, "Viroid Life: On Machines, Technics and Evolution," in *Deleuze and Philosophy: The Difference Engineer*, ed. Keith Ansell Pearson (London and New York: Routledge, 1997), 185).

20. Deleuze and Guattari, *A Thousand Plateaus*, 308.

21. Ibid., 275.

22. Ibid., 238.

23. Ibid., 176.

24. Ibid., 294.

25. Ibid., 295.

26. Ibid., 237.

27. Ibid., 241–242.

28. Braidotti, "Becoming Woman," 48.

29. Ansell Pearson, "Viroid Life."

30. Deleuze and Guattari, *A Thousand Plateaus*, 305.

31. Deleuze and Guattari, *A Thousand Plateaus*, 304.

32. Gilbert Simondon, "The Genesis of the Individuation," in *Incorporations*, ed. Jonathan Crary and Sanford Kwinter (Zhone Books, 1992), 313.

33. For a detailed analysis of Simondon's transduction see Adrian Mackenzie, *Transductions: Bodies and Machines at Speed,* (London and New York: Continuum, 2002).

34. Simondon "The Genesis of the Individuation," 313.

35. Simondon "The Genesis," 314.

36. Don Ihde, "Technology and Prognostic Predicaments," *AI and Society* 13 (1999): 44–51; Bernard Stiegler, *Technics and Time, 1: The Fault of Epimetheus*, trans. Richard Beardsworth and George Collins (Stanford: Stanford University Press, 1998).

37. Deleuze and Guattari, *A Thousand Plateaus*, 293.

38. See Robert Clowes, "Thinking in the Cloud: The Cognitive Incorporation of Cloud-Based Technology," *Philosophy and Technology* 28 (2015): 261–296.

39. Deleuze and Guattari, *A Thousand Plateaus*, 305.

40. Cf. Grosz's discussion on the different meanings of difference and the definition of becoming as one of these meanings. She reads Bergson's notion of intuition as a philosophical method that "occurs only when old and familiar methods by which intelligence always seeks to address the present and the new exhaust themselves and provide only generalizations rather than a concept uniquely suited to its object" (Elizabeth Grosz, "Bergson, Deleuze and the Becoming of Unbecoming," *Parallax* 11, no. 2 (2005): 7–8). The cell phone provides this unique moment that erodes the familiar dualism of mobile-immobile.

41. Gijs Mom, "Hop on the bus, Gus," *Transfers* 1, no. 1 (2011): 1–13.

42. Heike Weber, "Mobile Electronic Media: Mobility History at the Intersection of Transport and Media History," *Transfers*, 2011: 27–51.

43. Weber, "Mobile Electronic Media," 28.

44. Another concept that New Mobility Studies term is motility, which denotes "the capacity of entities (e.g., goods, information or persons) to be mobile in social and geographic space, or as the way in which entities access and appropriate the capacity for socio-spatial mobility according to their circumstances" (Vincent Kaufmann, Manfred Max Bergman, and Dominique Joye, "Motility: Mobility as Capital," *International Journal of Urban and Regional Research* 28, no. 4 (December 2004): 749–750). Motility refers to the intersection of geographic and *social* changes—it seeks to explain how mobility transforms societies. It is used to stress the social effects of mobility. Like virtual mobility, it is somewhat limited in its scope and hence cannot fully conceptualize the multiple aspects of becoming-mobile.

45. Also known as nomadicism (see Clair Colebrook, "Nomadicism," and Eugine Holland, "Nomadicism + Citizenship," both in *The Deleuze Dictionary*, ed. Adrian Parr (Edinburgh: Edinburgh University Press, 2005)).

46. Deleuze and Guattari, *A Thousand Plateaus*, 159.

47. Ibid., 381.

48. Ibid., 475.

49. Holland, "Nomadicism + Citizenship," 183.

50. Colebrook, "Nomadicism," 182.

51. Rosi Braidotti, *Patterns of Dissonance: A Study of Women and Conemporary Philosophy* (New York: Routledge, 1991), 111.

52. Deleuze and Guattari, *A Thousand Plateaus*, 361–363.

53. Ibid., 305.

54. When wireless transmission is affected due to weather conditions or to being inside a basement, one may experience voice distortion that reveals the "digitality" of cellular voice service.

55. Deleuze and Guattari, *A Thousand Plateaus*, 233.

56. For the ever-intensifying technological mediation of being-with-others see Yoni Van Den Eede, "In Between Us: On the Transparency and Opacity of Technological Mediation," *Foundations in Science* 16 (2011): 139–159.

57. Grosz explains: "Becoming is thus not a capacity inherited by life, an evolutionary outcome or consequence, but is the very principle of matter itself, with its possibilities of linkage with the living, with its possibilities of mutual transformation, with its inherent and unstable volatility" (Grosz, "Bergson," 10).

Bibliography

Ansell Pearson, Keith. "Viroid Life: On Machines, Technics and Evolution." In *Deleuze and Philosophy: The Difference Engineer*, by Keith Ansell Pearson, 180–210. London and New York: Routledge, 1997.

Armitage, John. "Beyond Postmodernism? Paul Virilio's Hypermodern Cultural Theory." Edited by Arthur Kroker and Marilouise Kroker. *CTheory* www.ctheory.net/articles.aspx?id=133 (2000).

Atterton, Peter. "Levinas and Our Moral Responsibility Toward Other Animals." *Inquiry* 54, no. 6 (2011): 622–649.

Aydin, Ciano. "The Artifactual Mind: Overcoming the 'Inside–Outside' Dualism in the Extended Mind Thesis and Recognizing the Technological Dimension of Cognition." *Phenomenology and the Cognitive Sciences*, 2013: 1–22.

Barnet, Belinda. "The Technical Evolution of Vannevar Bush's Memex." *Digital Humanities Quarterly* 2, no. 1 (2008).

Bassala, George. *The Evolution of Technology*. New York: Cambridge University Press, 1988.

Bogost, Ian. *Alien Phenomenology or What It's Like to be a Thing*. Minneapolis: University of Minnesota Press, 2010.

———. *The Geek's Chihuahua: Living with Apple*. Minneapolis: University Of Minnesota Press, 2015.

Braidotti, Rosi. *Nomadic Theory: The Portable Rosi Braidotti*. New York, Chichester, West Sussex: Columbia University Press, 2011.

Bush, Vannevar. "As We May Think." *The Atlantic Monthly*, July 1945.

Berardi, Franco "Bifo." *The Soul at Work: From Alienatio to Autonomy*. Translated by Francesca Cadel and Giuseppina Mecchia. Los Angeles: Semiotext(e), 2009.

Butler, Judith. *Gender Trouble: Feminism and the Subversion of Identity*. New York and London: Routledge, 1990.

Castells, Manuel. *The Information Age: Economy, Society and Culture, Vol I: The Rise of the Network Society*. Malden, Oxford, Vicotria: Blackwell Publishing, 1996, 2000.

Castells, Manuel, Mireia Fernandez-Ardevol, Jack Linchuan Qiu, and Araba Sey. *The Mobile Communication Society: A Cross-Cultural Analysis of Available Evidence on the Social Uses of Wirelss Communication Technology.* A research report prepared for the International Workshop on Wireless Communication Policies and Prospects, Annenberg School for Communication, University of Southern California, Los Angeles: Annenberg Research Network on International Communication, 2004.

Clark, Andy, and David Chalmers. "The extended mind." *Analysis* 58, no. 1 (1998): 7–19.

Clowes, Robert. "Thinking in the Cloud: The Cognitive Incorporation of Cloud-Based Technology." *Philosophy and Technology* 28 (2015): 261–296.

De Beistegui, Miguel. *The New Heidegger.* London and New York: Continuum, 2005.

De Certeau, Michel. *The Practice of Everyday Life.* Translated by Steven Rendall. Berkeley and Los Angeles: University of California Press, 1984.

De Preester, Helena. "Technology and the Body: The (Im)Possibilities of Re-embodiment." *Foundations of Science* 16 (2011): 119–137.

de Souza e Silva, Adriana. "From Cyber to Hybrid: Mobile Technologies as Interfaces of Hybrid Spaces." *Space and Culture* 9, no. 3 (August 2006): 261–278.

de Souza e Silva, Adriana, and Jordan Frith. "Locative Mobile Social Networks: Mapping Communication and Location in Urban Spaces." *Mobilities* 5, no. 4 (November 2010): 485–506.

Deleuze, Gilles, and Felix Guattari. *A Thousand Plateaus: Capitalism and Schizophrenia.* Translated by Brian Massumi. Minneapolis, London: University of Minnesota Press, 1987.

Derrida, Jacques. "Violence and Metaphysics: An Essay on the Thought of Emmanuel Levinas." In *Writing and Difference*, by Jacques Derrida. Translated by Alan Bass, 79–153. Chicago: University of Chicago, 1980.

———. "Geschlecht II: Heidegger's Hand." In *Deconstruction and Philosophy: The Texts of Jacques Derrida*, by John Sallis. Translated by John P. Leavey Jr., 161–196. Chicago and London: University of Chicago Press, 1989.

———. *Specters of Marx: The State of the Debt, The Work of Mourning & the New International.* New York, Oxon: Routledge, 1994.

———. *Adieu to Emmanuel Levinas.* Stanford: Stanford University Press, 1999.

Dodd, James. *Crisis and Reflection: An Essay on Husserl's Crisis of the European Sciences.* Dordrecht: Kluwer Academic Publishing, 2004.

Dormann, Claire, and Robert Biddle. "A Geo-Wiki Game for Urban Exploration and Community Participation." *Workshop: Exploring the Practices & Technologies of Representing the Urban Environment.* San Jose California, 2007.

Dreyfus, Hubert. "Beyond Hermeneutics: Interpretation in Late Heidegger and Recent Foucault." In *Hermeneutics: Questions and Prospects*, by Gary Shapiro and Alan Sica, 66–83. Amherst: University of Massachussets Press, 1984.

Dyer-Witheford, Nick. *Cyber-Marx: Cycles and Circuits of Struggle in High-Technology Capitalism.* Urbana and Chicago: University of Illinois Press, 1999.

Eaglestone, Robert. *Ethical Criticism: Reading After Levinas.* Edinburgh: Edinburgh Univeristy Press, 1998.

Engels, Frederick. "Dialectics of Nature." In *Karl Marx and Friedrich Engels Collected Works (Vol 25) - Friedrich Engels: Anti Duhring and Dialectics of Nature*. Translated by Clemens Dutt. International Publishers, 1987.

Farman, Jason. *Mobile Interface Theory: Embodied Space and Locative Media.* New York and London: Routledge, 2012B.

———. "Mobile Media Performances as Asynchronous Embodiment." *The International Hournal of Screendance* 2 (Spring 2012): 48–51.

Featherstone, Mike. "Ubiquitous Media: An Introduction." *Theory, Culture & Society* 26, no. 2–3 (2009): 1–22.

Feenberg, Andrew. *Questioning Technology*. London and New York: Routledge, 1999.

Ferraris, Maurizio. "Where are You? Mobile Ontology." *Seeing, Understanding, Learning in the Mobile Age.* Budapest: http://www.fil.hu/mobil/2005/Ferraris.pdf (accessed 22 October 2010), 2005.

Foucault, Michel. "Nietzsche, Genealogy, History." In *Language, Counter-memory, Practices: Selected Essays and Interviews*, by D.F. Bouchard, 139–164. Ithaca: Cornell University Press, 1977.

Friedberg, Anne. *The Virtual Window: From Alberti to Microsoft.* Cambridge, MA and London, England: The MIT Press, 2006.

Geser, Hans. *Towards a Sociological Theory of the Mobile Phone*. 3.0. http://socio.ch/mobile/t_geser1.pdf. 2004.

Glotz, Peter, Stefan Bertschi, and Chris Locke. *Thumb Culture: The Meaning of Mobile Phone for Society.* Bielfeld: Transcript Verlag, 2005.

Goggin, Gerard. *Cell Phone Culture: Mobile Technology in Everyday Life*. London and New York: Routledge, 2006.

———. "Adapting the Mobile Phone: The iPhone and its Consumption." *Continuum: Journal of Media & Cultural Studies* 23, no. 2 (2009): 231–244.

Green, Nicola. "Mobility, Memory and Identity." In *Mobile Technologies: From Telecommunications to Media*, by Gerard Goggin and Larissa Hjorth, 266–281. New York, London: Routledge, 2009.

Forss, Anette. "Cells and the (imaginary) Patient: The Multistable Practitioner–Technology–Cell Interface in the Cytology Laboratory." *Medicine, Health Care and Philosophy* 15, no. 3 (August 2012): 295–308.

Haddon, Leslie. "Domestication and Mobile Telephony." In *Machines That Become Us: The Social Context of Personal Communication Technology*, edited by James E. Katz, 43–55. New Brunswick and London: Transaction Publishers, 2001.

———. "Cultural Differences in Communication: Examining Patterns of Daily Life." *Mobile Communication and Social Change.* Seoul (Korea), 2004.

Handleman, Susan A. *Fragments of Redemption: Jewish Thought and Literary Theory in Benjamin, Scholem, and Levinas.* Bloomington and Indianapolis: Indiana University Press, 1991.

Hansen, Mark. "The Time of Affect, or Bearing Witness to Life." *Critical Inquiry* 30 (spring 2004B): 584–626.

Harman, Graham. *Tool-Being: Heidegger and the Metaphysics of Objects*. Peru, IL: Open Court Publishing, 2002.

Harper, Richard, and Lynne Hamill. "Kids Will be Kids: The Role of Mobiles in Teenage Life." *Mobile World*, 2005: 61–74.

Hauben, Jay. *Vannevar Bush and JCR Licklider: Libraries of the Future* 1945–1965. http://www.columbia.edu/~hauben/lof-final05.doc. 2004.

Hayles, N. Katherine. "Flesh and Metal: Reconfiguring the Mindbody in Virtual Environments." *Configurations,* 10 (2002): 297–320.

———. "Unfinished Work: From Cyborg to Cognisphere." *Theory Culture Society* 23, no. 7–8 (2006): 159–166.

———. "Hyper and Deep Attention: The Generational Divide in Cognitive Modes." *Profession,* (2007): 187–199.

Heidegger, Martin. *What is Called Thinking?* Translated by Glenn J. Gray and Fred D. Wieck. New York: Harper & Row, 1968.

———. "The Question Concerning Technology." In *The Question Concerning Technology, and Other Essays*. Translated by W. Lovitt. New York: Garland Pub., 1977.

———. *Being and Time.* Translated by Juan Stambaugh. Albany: State University of New York Press, 1996.

Hoskins, Andrew. "Media, Memory, Metaphor: Remembering and the Connective Turn." *Parallax* 17, no. 4 (2011): 19–31.

Ihde, Don. *Technics and Praxis: A Philosophy of Technology.* Dordrecht: Reidel Publishing Company, 1979.

———. *Consequences of Phenomenology.* Albany: State University of New York Press, 1986.

———. *Technology and the Lifeworld: From Garden to Earth.* Bloomington and Indianapolis: Indiana University Press, 1990.

———. *Postphenomenology: Essays in Postmodern Context.* Evanstone: Northwestern University Press, 1995.

———. *Expanding Hermeneuics: Visualism in Science.* Evanstone (IL): Northwestern University Press, 1998B.

———. "Technology and Prognostic Predicaments." *AI and Society* 13 (1999): 44–51.

———. *Bodies in Technology.* Minneapolis and London (UK): University of Minnesota Press, 2002.

———. "A Response to My Critics." *Techné* 7 (2003): 110–117.

———. *Postphenomenology - Again?* Working Papers from Centre for STS Studies, University of Aarhus, The Centre for STS Studies, 2003D.

———. "Aging: I don't want to be a cyborg!" *Phenomenology and the Cognitive Sciences* 7, no. 3 (2008): 397–404.

———. *Postphenomenology and Technoscience: The Peking University Lectures.* New York: State University of New York Press, 2009.

———. *Heidegger's Technologies: Postphenomenological Perspectives.* New York: Fordham University Press, 2010.

———. *Experimental Phenomenology: Multistabilities (second edition).* Albany: State University of New York Press, 2012.

Introna, Lucas D., and Fernando M. Ilharco. "Ontological Screening of Contemporary Life." *European Journal of Information Systems* 13 (2004): 221–234.

Irwin, Stacy. "Technological Other/Quasi Other: Reflection on Lived Experience." *Human Studies* 28 (2006): 453–467.

Jonas, Hans. "Towards a Philosophy of Technology." In *Philosophy of Technology: The Technological Condition*, by Robert C. Scharff and Val Dusek, 191–205. Malden, Oxford, Victoria: Blackwell Publishing, 2003.

Katz, James. *Magic in the Air: Mobile Communication and the Transformation of Social Life*. New Brunswick and London: Transaction Publishers, 2006.

Kaufmann, Vincent, Manfred Max Bergman, and Dominique Joye. "Motility: Mobility as Capital." *International Journal of Urban and Regional Research* 28, no. 4 (December 2004): 745–756.

Kline, Stephen J. "What is Technology." In *Philosophy of Technology: The Technological Condition, An Anthology*, by Robert C. Scharff and Val Dusek, 210–212. Malden, Oxford, Victoria: Blackwell Publishing, 2003.

Koopman, Colin. *Genealogy As Critique: Foucault and the Problems of Modernity*. Bloomington: Indiana University Press, 2013.

Lash, Scott. *Critique of Information*. London: Sage, 2002.

Latour, Bruno. "On Recalling ANT." In *Actor Network Theory and After*, by John Law and John Hassard, 15–25. Oxford and Malden: Blackwell Publishing, 1999.

———. *Reassembling the Social: An Introduction to Actor-Network Theory*. Oxford and New York: Oxford University Press, 2005.

Levinas, Emmanuel. *Totality and Infinity: An Essay on Exteriority*. Pittsburgh: Duquense University Press, 1995.

Levinson, Paul. *Cell phone: The Story of the World's Most Mobile Medium and How It Has Transformed Everything!* New York and Basingstoke: Palgrave MacMillan, 2004.

Licklider, JCR. *Libraries of the Future*. Cambridge: The MIT Press, 1965.

Ling, Richard. "The social juxtaposition of mobile telephone conversations and public spaces." *The Social Consequences of Mobile Telephones*. Chunchon, Korea, 2002.

———. *The Mobile Connection: The Cell Phone's Impact on Society*. Morgan Kaufman, 2004.

———. *New Tech, New Ties: How Mobile Communication Is Reshaping Social Cohesion*. Cambridge and London: The MIT Press, 2008.

Ling, Richard, and Leslie Haddon. "Machines that Become Us." In *Mobile Telephone, Mobility and the Coordination of Everyday Life*, by James E. Katz. New Brunswick: Transaction, 2003.

Llewelyn, John. *Appositions of Jacques Derrida and Emmanuel Levinas*. Bloomington and Indianapolis: Indiana University Press, 2002.

Mackenzie, Adrian. *Transductions: Bodies and Machines at Speed*. London and New York: Continuum, 2002.

MacKenzie, Donald. "Marx and the Machine." In *Knowing Machines: Essays on Technical Change*, by Donald MacKenzie, 23–48. Cambridge and London: The MIT Press, 1996.

Manovich, Lev. *The Language of New Media*. Cambridge and London: The MIT Press, 2001.

———. "The Poetics of Augmented Space." *Visual Communications* (SAGE) 5, no. 2 (2006): 219–240.

Marx, Karl. *Grundisse: Foundations of the Critique of Political Economy*. London: Penguin, 1973.

———. *Capital: A Critique of Political Economy*. Edited by Fredrick Engels. Translated by Samuel Moore and Edward Aveling. *Vol. 1: The Process of Capitalist Production*. International Publishers, 1983.

Marx, Karl, and Friedrich Engel. *The German Ideology*. International Publishers Co, 1970.

Marx, Leo. "Technology: The Emergence of a Hazardous Concept." In *Technology and the Rest of Culture*, by Arien Mack, 23–46. Ohio State University Press, 2001.

May, Harvey, and Greg Hearn. "The Mobile Phone as Media." *International Journal of Cultural Studies* 8, no. 2 (2005): 195–211.

Mayer-Schönberger, Viktor. *Delete: The Virtue of Forgetting in the Digital Age*. Princeton: Princeton University Press, 2011.

Mol, Annemarie. "Ontological Politics. A Word and Some Questions." In *Actor Network Theory and After*, by John Law and John Hassard, 74–89. Oxford and Malden: Blackwell Publishing, 1999.

Mom, Gijs. "Hop on the bus, Gus'." *Transfers* 1, no. 1 (2011): 1–13.

Negroponte, Nicholas. *Being Digital*. New York: Vintage Books (Random House), 1996.

O'Gorman, Marcel. "Taking Care of Digital Dementia." *CTheory*, 2015.

Okabe, Daisuke. "Emergent Social Practices, Situations and Relations through Everyday Camera Phone Use." *Mobile Communication and Social Change*. Seoul, Korea, 2004.

Olkowski, Dorothea. "Flows of Desire and the Body-Becoming." In *Becomings: Explorations in Time, Memory, and Futures*, by Elizabeth Grosz, 98–116. Ithaca: Cornell University Press, 1999.

Peperzak, Adriaan Theodoor. *Beyond: The Philosophy of Emmanuel Levinas*. Evanston: Northwestern University Press, 1999.

Pinchot, Jamie, Karen Paullet, and Daniel Rota. "How Mobile Technology is Changing Our Culture." *Journal of Information Systems Applied Research* 4, no. 1 (2011): 39.

Plant, Sadie. *On the Mobile: The Effects of Mobile Telephones on Social and Individual Life*. Motorola, 2003, http://www.motorola.com/mot/documents/0,1028,333,00.pdf.

———. *Zeros + Ones: Digital Women + The New Technoculture*. New York: Doubleday, 1997.

Postman, Neil. *Technolopoly: The Surrender of Culture to Technology*. New York: Randon House, 1992.

Reading, Anna. "Seeing Red: A Political Economy of Digital Memory." *Media, Culture & Society* 36, no. 6 (2014): 748–760.

Rheingold, Howard. *Tools for Thought: The History and Future of Mind-Expanding Technology*. Cambridge: MIT Press, 2000.

Richardson, Ingrid. "Faces, Interfaces, Screens: Relational Ontologies of Framing, Attention and Distraction." *Transformation* 18 (The Face and Technology) (2010): http://www.transformationsjournal.org/journal/issue_18/article_05.shtml.

———. "Mobile Technosoma: Some Phenomenological Reflections on Itinerant Media Devices." *Fibreculture*, 2005: http://journal.fibreculture.org/issue6/.

Riis, Søren. "Towards the Origin of Modern Technology: Reconfiguring Martin Hei-Degger's Thinking." *Continental Philosophy Review* 44 (2011): 103–117.

Riviere, Carole. "Mobile Camera Phones: A New Form of 'Being Together' in Daily Interpersonal Communication." In *Mobile Communications: Re-negotiation of the Social Sphere*, by Richard Seyler Ling and Per E. Pederson, 167–185. London: Springer, 2005.

Rosenberger, Robert. "Perceiving Other Planets: Bodily Experience, Interpretation, and the Mars Orbiter Camera." *Human Studies*, no. 31 (2008): 63–75.

———. "A Case Study in the Applied Philosophy of Imaging: The Synaptic Vesicle Debate." *Science Technology and Human Values* 36, no. 1 (2011): 6–32.

———. "Embodied Technology and the Dangers of Using the Phone while Driving." *Phenomenology and the Cognitive Sciences* 11, no. 1 (March 2012): 79–94.

Sandberg, Anders, and Nick Bostrom. "Converging Cognitive Enhancements." *Annals of the New York Academy of Sciences* 1093, no. 1 (2006): 201–227.

Scharff, Robert C., and Val Dusek. *Philosophy of Technology: The Technological Condition, An Anthology*. Malden, Oxford, Victoria: Blackwell, 2003.

Scifo, Barbara. "The Domestication of Camera-Phone and MMS Communication: The Early Experiences of Young Italians." In *A Sense of Place: The Global and the Local in Mobile Communication*, by K. Nyiri, 363–74. Vienna: Passagen Verlag, 2003.

———. "The Sociocultural Forms of Mobile Personal Photographs in a Cross-Media Ecology: Reflections Starting from the Young Italian Experience." *Knowledge, Technology and Policy* 22, no. 3 (2009): 185–194.

Silverstone, Roger, and Knut H. Sorensen. "Towards the 'Communication Society'." In *Media, Technology and Everyday Life in Europe: From Information to Communication*. Hants and Burlington: Ashgate, 2005.

Silverstone, Roger, Eric Hirsch, and David Morley. "Information and Communication Technologies and the Moral Economy of the Household." In *Consuming Technologies: Media and Information in Domestic Spaces*, by Roger Silverstone, 15–31. London: Routledge, 1992.

Simondon, Gilbert. "The Genesis of the Individuation." In *Incorporations*, by Jonathan Crary and Sanford Kwinter, 297–319. Zhone Books, 1992.

Stiegler, Bernard. *Technics and Time, 1: The Fault of Epimetheus*. Translated by Richard Beardsworth and George Collins. Stanford: Stanford University Press, 1998.

———. "Our Ailing Educational Institutions." *Culture Machine* 5 (2003).

———. "Anamnesis and Hypomnesis: The Memories of Desire." In *Technicity*, by Louis Armand and Arthur Bradley, 15–41. Prague: Litteraria Pragensia, 2006.

———. *Technics and Time, 2: Disorientation*. Translated by Stephen Barker. Stanford: Stanford University Press, 2009.

———. "The Carnival of the New Screen: From Hegemony to Isonomy." In *The YouTube Reader*, by Pelle Snickars and Patrick Vonderau, 40–59. Stockholm: National Library of Sweden, 2009B.

———. *Taking Care of Youth and Generations*. Stanford: Stanford University Press, 2010.

———. *For a New Critique of Political Economy*. Polity, 2010B.

Sturdevant, Rick W. "NAVSTAR, the Global Positioning System: A Sampling of Its Military, Civil, and Commercial Impact." In *Societal Impact of Spaceflight*, by Steven J. Dick and Roger D. Launius, 331–351. Government Printing Office, 2009.

Townsend, Anthony M. "Life in the Realtime City: Mobile Telephones and Urban Metabolism." *Journal of Urban Technology* 7, no. 2 (2000): 85–104.

Turkle, Sherry. *Alone Together: Why We Expect More from Technology and Less from Each Other*. Basic Books, 2012.

Van Den Eede, Yoni. "In Between Us: On the Transparency and Opacity of Technological Mediation." *Foundations in Science* 16 (2011): 139–159.

van de Poel, Ibo, and Peter-Paul Verbeek. "Editorial: Ethics and Engineering Design." *Science Technology and Human Values* 31, no. 3 (2006): 223–236.

Vanderbeeken, Robrecht. "The Screen as an In-between." *Foundations of Science* 16 (2011): 245–257.

Verbeek, Peter-Paul. *What Things Do: Philosophical Reflections on Technology, Agency and Design*. University Park, PA: The Pennsylvania State University Press, 2005.

———. "Materializing Morality: Design Ethics and Technological Mediation." *Science Technology and Human Values* 31, no. 3 (2006): 361–380.

———. "Cyborg Intentionality: Rethinking the Phenomenology of Human–Technology Relations." *Phenomenology and Cognitive Science* 7 (2008): 387–395.

———. "Cultivating Humanity: Toward a Non-Humanist Ethics of Technology." In *New Waves in Philosophy of Technology*, by Jan Kyrre Berg Olsen, Evan Selinger and Soren Riis, 241–266. Basingstoke: Palgrave Macmillan, 2009.

Vincent, Jane, Leslie Haddon, and Lynne Hamill. "The Influence of Mobile Phone Users on the Design of 3G Products and Services." *The Journal of the Communications Network* 4, no. 4 (October–December 2005): 69–73.

Virilio, Paul. *The Vision Machine*. Bloomington & Indianapolis: Indiana University Press, 1994.

Waldenfels, Bernhard. "Levinas and the face of the other." In *The Cambridge Companion to Levinas*, by Simon Critchley and Robert Bernasconi, 63–81. Cambridge University Press, 2004.

Weber, Heike. "Mobile Electronic Media: Mobility History at the Intersection of Transport and Media History." *Transfers*, 2011: 27–51.

Wellner, Galit. "Wall-Window-Screen: How the Cell Phone Mediates a Worldview for Us." *Humanities and Technology Review* (2011): 77–104.

———. "Multi-Attention and the Horcrux Logic: Justifications for Talking on the Cell Phone While Driving." *Techné Research in Philosophy and Technology* 18, no. 1/2 (2014): 48–73.

Wendling, Amy E. *Karl Marx on Technology and Alienation*. Basingstoke: Palgrave Macmillan, 2011.

Index

www.ingramcontent.com/pod-product-compliance
Lightning Source LLC
LaVergne TN
LVHW092132260926
848997LV00001B/3
9780739198483